FORESTS & WATER

GUIDELINES
THIRD EDITION

LONDON : HMSO

© *Crown copyright 1993*
Applications for reproduction should be made to HMSO
First edition Crown copyright 1988
Second edition Crown copyright 1991

ISBN 0 11 710321 7

Enquiries relating to this publication should be addressed to:
The Research Publications Officer,
The Forestry Authority,
Research Division,
Alice Holt Lodge,
Wrecclesham,
Farnham,
Surrey
GU10 4LH

Printed in the United Kingdom for HMSO.
Dd.0294480, C50, 11/93, 3396/4, 5673, 258903.

FORESTS & WATER GUIDELINES

CONTENTS

FOREWORD	iv
INTRODUCTION	**1**
Background	1
Planning and managing operations	1
Water and the law	1
Grant and felling licence applications	2
Liaison between forestry and water interests	2
CATCHMENT WATER PATHWAYS	**3**
The adjacent land	3
The riparian zone	4
The aquatic zone	4
THE USE OF WATER	**5**
Principal uses	5
Requirements of water users	5
Requirements of wildlife	5
Requirements of fish	6
Potential problems	6
SPECIFIC RECOMMENDATIONS ON FOREST OPERATIONS	**11**
Critical loads, catchment assessments and ameliorative treatments	11
Ground preparation	11
Buffer areas	13
Managing riparian vegetation	14
Road construction and maintenance	15
Harvesting	16
Pesticides	17
Fertilisers	18
Storage and handling of chemicals and fuel oils	18
Contingency plan in case of chemical or fuel oil spillage	18
Ponds for fire-fighting or wildlife	18
WORKING CHECKLIST	**19**
Appendix I: Sources of advice	21
Appendix II: The water industry and water legislation	25
Appendix III: The forestry industry	27
Glossary	28
ACKNOWLEDGEMENTS	**31**
BIBLIOGRAPHY	**32**

FORESTS AND WATER GUIDELINES

FOREWORD

This is the third edition of Forests and water guidelines. The recommendations contained in these guidelines result from extensive consultation throughout the water and forestry industries – with universities and research institutes, government departments and agencies, and other organisations concerned with these matters. Comments have been considered by a core group with representation from both government departments and other public bodies with regulatory responsibilities for forestry and water. The objective has been to make the guidelines reflect the results of the most recent research and experience. These guidelines are based on the consensus view of the scientific issues, informed by a knowledge of practical implementation.

R T Bradley
Head of the Forestry Authority
Forestry Commission

INTRODUCTION

Forests and woodlands, under sensitive management, can substantially enhance the environment, including the freshwater ecosystem. The Forestry Authority has published a series of guidelines which seek to ensure that forests and woodlands are managed and developed so that their social, environmental and economic benefits are realised for the community at large, in line with Government policy. The series now covers Forests and Water, Forest Landscape Design, Lowland Landscape Design, Forest Nature Conservation, Community Woodland Design, and Forest Recreation.

Most forest operations have potential effects on the quantity and quality of water moving through forest stands. *Forests and water guidelines* advises owners and managers how woodlands and forests influence the freshwater ecosystem and gives guidance on how operations should be carried out in order to protect and enhance the water environment. The guidelines apply equally to Forest Enterprise and the private sector. Grant approvals and felling licences are subject to the standards set out in this book. A number of laws exist to protect water interests. Although *Forests and water guidelines* has no formal legal status, in the event of a prosecution, failure to follow the guidelines is likely to adversely affect the position of the forest owner.

Until recently the creation of forests has been primarily in the uplands. **In future more forests are likely to be established in the lowlands, including close to towns and cities.** New semi-natural woodlands are being encouraged through initiatives like the new native pinewood grants. The relationship between forest and water differs from the uplands to the lowlands and between forest types, and this edition of *Forests and water guidelines* addresses both traditional and new developments. Finally, restocking after felling of the first rotation presents an important opportunity for restructuring a significant area of upland forestry in line with current forestry objectives and management practice.

BACKGROUND

The first edition of *Forests and water guidelines* was published in 1988 following a 'water workshop' organised by the **Forestry Commission** and the **Water Research Centre** at York in 1986. The first edition contained the best guidance available at the time, while also informing the water industry about forest operations which affect their responsibilities. It was well received by the forestry and water industries alike.

Since 1987 extensive research has been conducted on the deposition and effects of acidic pollutants. In June 1990 the **Forestry Commission** and the **Department of the Environment** jointly sponsored a workshop to assemble an expert view on the role of forestry within these effects. The workshop report, *Forests and surface water acidification,* was published in 1991. As a result of this workshop the sections of the guidelines dealing with acidification were greatly expanded and a second edition of *Forests and water guidelines* was published in 1991.

This third edition draws on recent and continuing studies of the various effects of land use, pollutant inputs and forest operations.

PLANNING AND MANAGING OPERATIONS

The sensitivity of any water catchment to forest operations must be identified and taken into account at the planning stage.

Forest management can have profound effects on streams, lakes and reservoirs.

Good management of watercourses and adjacent areas enhances both the wildlife and scenic values of forests while improving conditions for tree growth and protecting forest roads.

Poor management, in contrast, can lead to increased soil and stream erosion, greater turbidity, sedimentation and pollution. These increase costs of treatment for drinking water, and can cause damage to roads, fisheries and wildlife. There can also be a loss of some forest soil and thus a reduction in potential productivity.

In sum, good water management is not merely avoiding harm, whether to water running through the forest or to downstream users. It requires active and imaginative measures. Successful water management is ecologically, aesthetically and, possibly, economically rewarding.

WATER AND THE LAW

Forest managers must meet their legal obligations under the Control of Pollution Act (1974) as amended by the Water Act (1989) (in Scotland), the Water Resources Act (1991) (in England and Wales) and the Food and Environment Protection Act (1985), and other relevant legislation, when carrying out all forestry operations. It is an offence under the Control of Pollution Act (1974) (in Scotland) and the Water Resources Act (1991) (in England & Wales) to cause or knowingly permit the entry of poisonous, noxious or polluting material into any watercourse or water body. Advice is available from the Forestry Authority. When in doubt forest managers should liaise in advance with the local office of their water regulatory authority or water undertaker. Consultation with the water regulatory authority is specifically required prior to aerial applications of fertilisers and of herbicides and pesticides on land adjacent to water, and to any application of sewage sludge to land.

Throughout these guidelines the term 'water regulatory authority' should, in Scotland, be taken to refer to the local river purification authority (RPA) – the seven river purification boards (RPBs) on the mainland, and the three Islands councils for their areas – and in England and Wales, to the National Rivers Authority (NRA). The term 'water undertaker' should, in Scotland, be taken to refer to the regional or Islands councils and, in England and Wales, to the water utility companies. In Scotland, the district salmon fisheries boards have statutory responsibilities for salmon and sea trout. In England and Wales, salmon and sea trout are included in the responsibilities of the National Rivers Authority for fisheries in general. Addresses are given in Appendix I.

Water legislation and the structure of the water industry are described in Appendix II.

GRANT AND FELLING LICENCE APPLICATIONS

The Forestry Authority gives grants under the Woodland Grant Scheme (WGS) for the establishment and management of woodlands. It is a condition of the WGS that 'you must meet the standards of environmental protection and practice set out in our published guidelines current at the time of approval'. In many cases the Forestry Authority will expect details to be given of how water issues will be dealt with in a particular woodland. However, these guidelines apply to all work included in the WGS and failure to abide by them can result in the withholding or reclaiming of grant. Further details are given in the booklet on the Woodland Grant Scheme, which is also in the WGS Applicant's Pack.

Plans of operations under the Dedication Scheme must also work to current standards.

Before giving permission to fell trees under a felling licence, the Forestry Authority will consider whether the standards of these guidelines will be met. More information can be found in the booklet *Tree felling – licences and permissions*.

- **Indicative forestry strategies** identify those areas where significant planting is to be preferred or where there are particular environmental sensitivities which must be addressed and satisfied before approval can be given for grant aid. They are drawn up by the local planning authority and were introduced in Scotland in 1990 and in England and Wales in 1992. In drawing up such strategies, the planning authorities consult with the river purification boards in Scotland and the National Rivers Authority in England and Wales.

- **Formal environmental assessment procedures** for new woodland were introduced in 1988. These procedures assess the likely effects which a proposed project may have on the environment, and allow the Forestry Authority to consider the effects, and ways in which they may be reduced, before making a decision on a grant application. Details are set out in the booklet *Environmental assessment of new woodlands*.

- **A public register of new planting applications** is held at local Forestry Authority offices (called conservancies). This allows people to find out about proposals for new woodlands in the local area and to make comments if they wish.

- **The Forestry Authority has agreed a process of consultation** with a number of statutory bodies before determining applications either under the WGS or for a felling licence. These arrangements are revised and agreed from time to time by the Forestry Authority and the statutory bodies. Most planting and felling proposals are uncontroversial; the regional advisory committees of the Forestry Authority have a conciliatory role in those few cases where problems do arise. Details are given in the Forestry Authority booklet *Consultation for grant and felling applications*.

These procedures, coupled with the strict environmental standards which are an integral part of the WGS, ensure that work in woodlands is properly planned and carried out. These Guidelines apply equally to woodlands managed by Forest Enterprise, and compliance with them is a condition of Forestry Authority approval of Forest Enterprise operations.

LIAISON BETWEEN FORESTRY AND WATER INTERESTS

Although the Forestry Authority has a formal process to take account of water interests, forest managers will benefit from establishing contact with the appropriate water regulatory authority and water undertaker and, when appropriate, seeking advice. Relevant addresses are given in Appendix I.

Water catchment zones

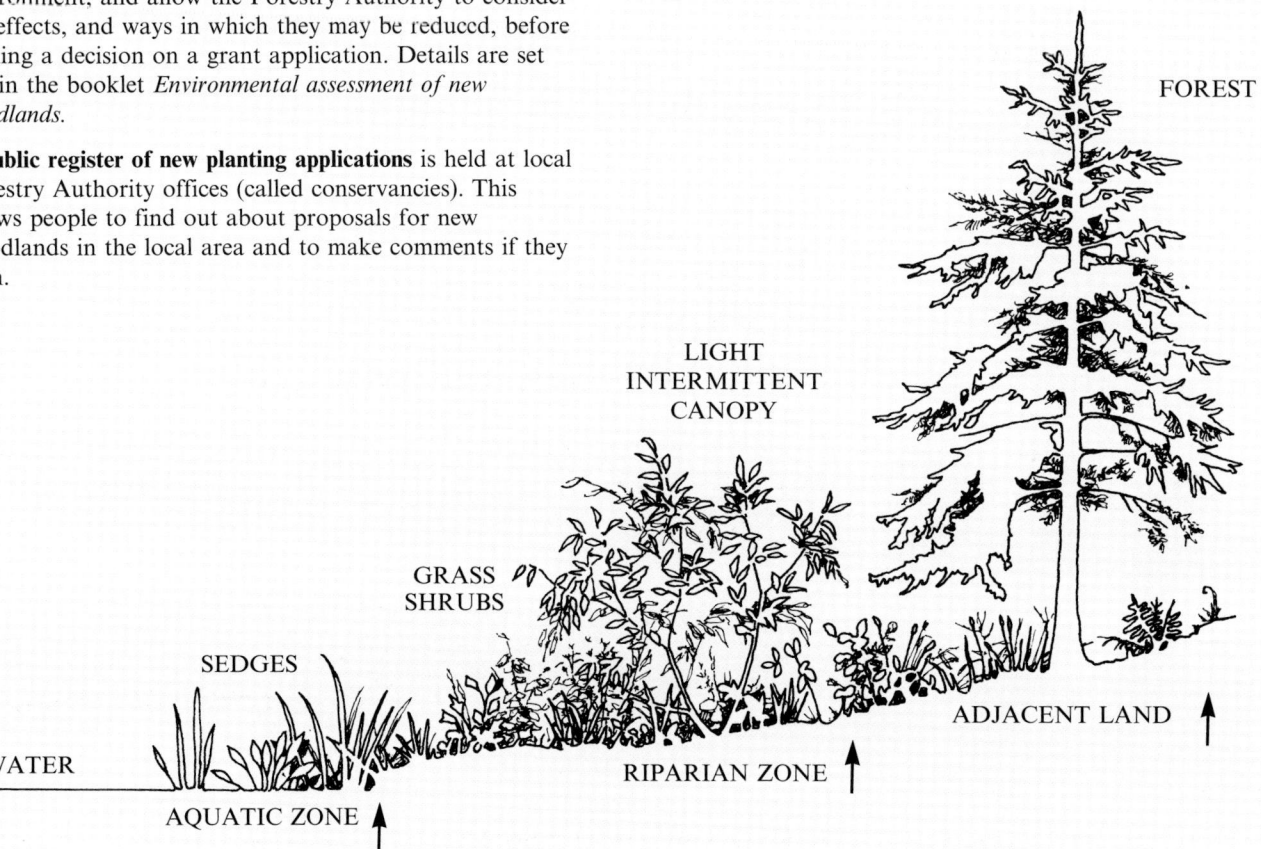

CATCHMENT WATER PATHWAYS

It is helpful to think of water moving through three contiguous zones within a catchment – the adjacent land, the riparian zone and the aquatic zone.

THE ADJACENT LAND

This is the largest zone and principal water gathering ground. Water enters a catchment as precipitation, and passes into the soil through the vegetation layer which can exert a strong influence on both the quantity and quality of the water. The quantity of precipitation reaching the soil as throughfall and stemflow is reduced by the interception loss – that is, the proportion retained by the vegetation layer and subsequently evaporated back to the atmosphere. Forests, particularly of conifers, have a higher interception loss than grassland due to their relative size and the surface roughness of their canopies. The quality of throughfall is also altered by the evaporative loss, as a result of the capture of mist, aerosols and pollutant gases, and of chemical interactions within the vegetation layer. Having passed through the vegetation layer and into the soil, water is taken up by vegetation and returned to the atmosphere through the process of transpiration, or is retained by the soil, or leaves the soil as drainage.

The amount of water following each of these routes is influenced by the nature of the vegetation and therefore by land use practices. For example, transpiration losses from forests are believed to be less than from grasslands, helping to offset the greater interception losses. Cultivation and harvesting will reduce evaporation and transpiration loss due to the removal of the vegetation canopy, and result in more water leaving the soil as drainage.

Drainage water can take a number of pathways over and through the soil and bedrock *en route* to the catchment outlet. The pathways taken will depend on topography, soil, drift deposit and underlying geology, and will have a marked influence on the timing, volume and quality of water reaching the aquatic zone. Steep slopes, poorly draining soils or shallow impermeable bedrock result in superficial pathways and a fast catchment response time to precipitation. Gentle slopes, freely draining soils, deep drifts and porous bedrock result in deeper pathways and a delayed and attenuated response. Superficial waters tend to be soft, brown and acidic, reflecting their short passage through the upper organic soil horizons. Waters following deeper pathways tend to be harder, clearer and more alkaline, due to the longer period in which they are in contact, and able to react, with soil and rock minerals.

Forest operations can alter drainage water pathways which, in turn, may have a significant effect upon water quality and quantity. For example, in some soil types ploughing and drainage favour surface pathways and lead to higher and quicker flows. In these circumstances soil disturbance can result in a marked increase in erosion and sedimentation. This may have profound consequences for water interests downstream, which may be difficult, expensive or impossible to put right. Legal action may follow the pollution of streams or the killing of fish.

The effects of poor practice may not always be readily apparent, particularly in catchments underlain by porous rocks which form aquifers for drinking water. For example, the careless use of pesticides can lead to the contamination of soil drainage which may not reach a river or borehole abstraction point for several decades. In some areas ground water is the sole source of both public and private water supply. Once contaminated, it may be difficult or impossible to restore. Pesticide applications within areas used as aquifers for public water supplies must, therefore, be very carefully planned.

Water pathways

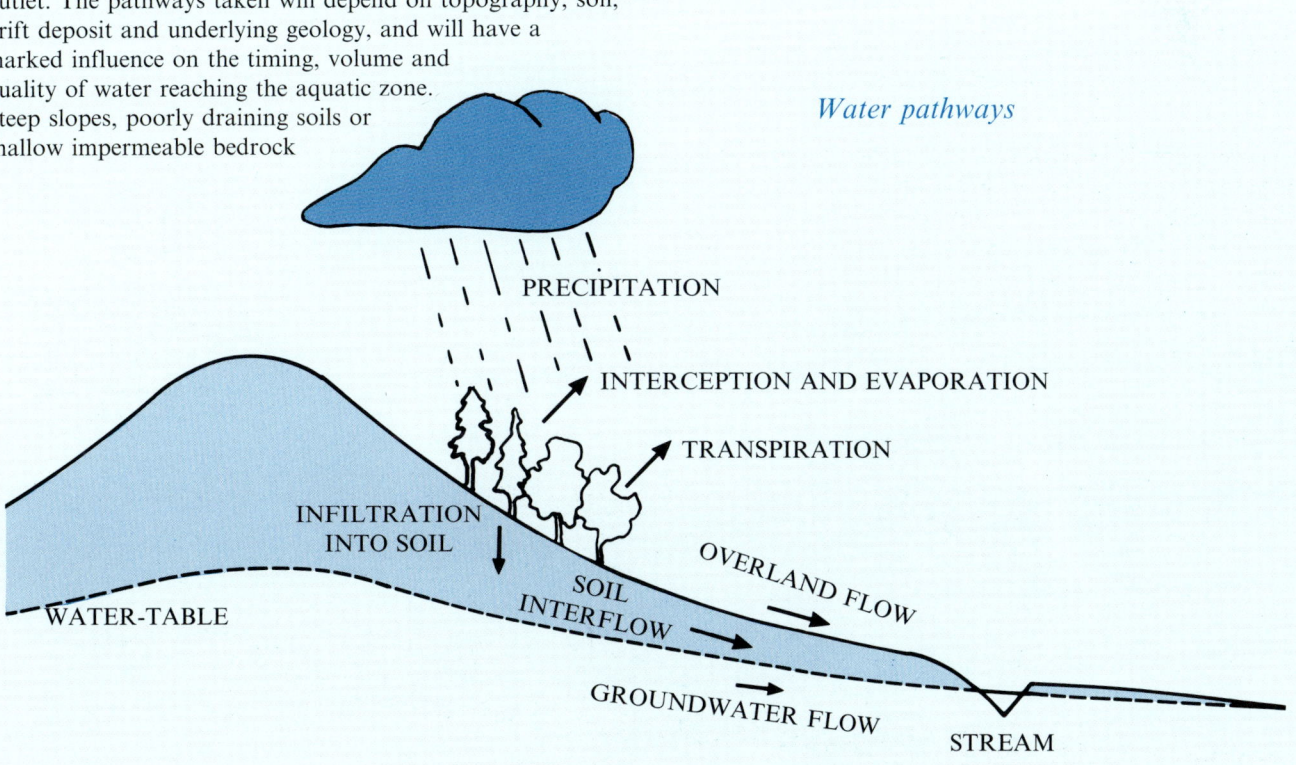

THE RIPARIAN ZONE

The riparian zone is the land immediately adjoining the aquatic zone and influenced by it. Some of the soils in this zone are often at or near to saturation, and it is here that water flowing through the soil and bedrock of the adjacent land may re-emerge to contribute to stream flow. Such inputs tend to be rich in minerals and, together with occasional contributions of silt and abundant moisture, provide particularly favourable growing conditions for plants. The zone is ecologically rich because of the variety of habitat types that are often present – such as stream banksides, flushes, areas of natural erosion and deposition, bog and wet woodland. The wetness of the soils and the characteristic instability of stream banks mean that the zone is very sensitive to disturbance. Management must aim to protect and encourage the diversity of these rich habitats for the benefit of the whole forest.

THE AQUATIC ZONE

The aquatic zone is the ground frequently or permanently under water, forming streams, rivers, ponds and lakes.

Upland waters are an important source of water for public supply, a large number of reservoirs having been built in the last 150 years. Much of this water requires only simple treatment, maybe limited to disinfection, with no need for basic filtration. Water from upland reservoirs is therefore relatively cheap, compared with supplies drawn from lowland rivers. It is very sensitive to disturbance and even limited deterioration can disrupt water treatment facilities, adding greatly to treatment costs. The uplands also contain a large number of small private water supplies which are at even greater risk from disturbance since they frequently undergo no form of treatment at all. Forests can also diminish the reliable yield of water making it necessary to seek alternative or supplementary supplies.

Upland waters are often valuable for fisheries, wildlife and recreation. Small headwater streams are particularly important as spawning habitat for salmonids, and particularly vulnerable to forest operations since such a large proportion of their catchment can be affected. The consequences are serious if they are filled or blocked off by sediment or debris from tree felling. Good management will protect the aquatic zone from such damage and seek to improve the habitat for fisheries and wildlife. For example, the careful siting of logs to create pools in fast flowing sections of stream can provide a range of depths and habitats for fish and other aquatic life. The construction of small groynes may help to stabilise gravel beds, although care is needed if serious bank erosion on the opposite side of watercourses is to be avoided. Forest managers are strongly advised to seek the advice of the water regulatory authority before undertaking such work.

Lowland waters tend to be poorer in quality as a result of agricultural, urban and industrial pollution. Forestry is a less intensive land use than agriculture, and the planting of woodlands on arable land in the lowlands is likely to have advantages in terms of water quality.

Edges

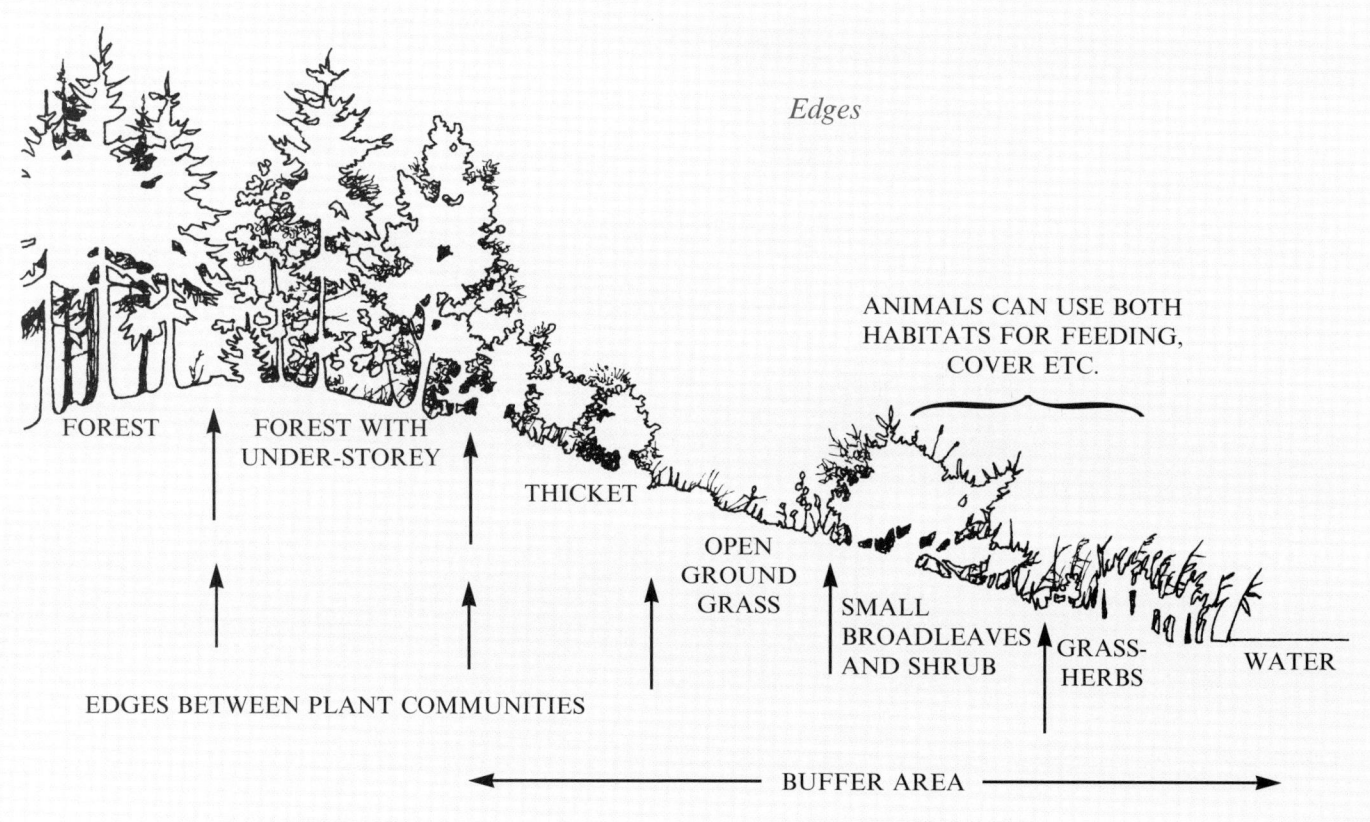

THE USE OF WATER

PRINCIPAL USES

Water in and flowing from forested catchments may be used for:

- wildlife, including recreational and commercial fisheries;
- abstraction for drinking purposes;
- hydro-electric power generation;
- agriculture and irrigation;
- industrial purposes;
- other recreational uses, like canoeing and sailing.

REQUIREMENTS OF WATER USERS

The requirements of water users vary considerably. Water supply and power generation need reliable yields; wildlife and recreational uses adjust to a particular flow and quality regime. If this is changed problems may occur, with possible increased costs for specific water users or the general public. The table below summarises the principal requirements of particular users. *Problems often arise in combination and may lead to secondary difficulties.* It is useful to concentrate first on the needs of aquatic wildlife because if these are met the water is likely to be suitable for most other users. Conservation of wildlife is also a general requirement under the Government's forestry policy.

REQUIREMENTS OF WILDLIFE

Aquatic wildlife will in general benefit from water conditions which vary within the natural limits appropriate to the site. Water chemistry, temperature, oxygenation, velocity and depth, and levels of suspended solids, sedimentation and pollutants may all be factors of importance.

Requirements are:

- **adequate light** reaching the water to support aquatic plants and algae and to maintain temperatures suitable for animal metabolism;
- **a diverse physical environment** appropriate to the site, such as pools; riffles; gravel bars; steep, shallow and undercut banks; wetlands; ponds and backwater channels; dry river terraces and alluvial flood-plains;
- **a range of appropriate vegetation,** such as algae on stony streambeds, rooted plants in the silt or sand of less turbulent waters, bankside vegetation which helps to control erosion and provide shelter, and a wide diversity in the riparian zone;
- **water free of contaminants** in harmful concentrations;
- **nutrient levels** appropriate to that water body.

REQUIREMENTS FOR VARIOUS USES OF WATER AND WATERCOURSES

Factor	Wildlife including fisheries	Drinking water supply	Hydro-electricity	Fish Farming	Irrigation	Land drainage/ flood control	Disposal and self-purification of waste (a)	Industrial supply (b)	Recreation
WATER QUANTITY									
High yield		✓	✓	✓	✓			✓	✓
Steady supply		✓	✓	✓	✓		✓	✓	✓
Peak flows controlled		✓	✓			✓			✓
Appropriate seasonal flows	✓	✓			✓		✓		✓
WATER QUALITY									
Water clarity	✓	✓		✓				✓	
Sediment free	✓	✓	✓	✓		✓		✓	
Un-coloured		✓						✓	
Untainted	✓	✓		✓				✓	
Low nutrient		✓						✓	
Free of toxic chemicals	✓	✓		✓	✓			✓	✓
Non-pathogenic	✓	✓		✓	✓			✓	✓
Moderate pH	✓	✓	✓	✓	✓		✓	✓	✓
Low algal production		✓	✓	✓	✓			✓	✓
OTHER REQUIREMENTS									
Adequate watercourse channels	✓					✓	✓		✓
Appropriate water temperature range	✓	✓		✓			✓		✓
Adequate oxygen level	✓	✓		✓			✓		✓
Adequate insect and algal production	✓								✓

(a) Poor upstream water quality could reduce the assimilatory capacity of the river downstream for legitimately discharged effluents.
(b) Requirements of industrial supplies vary, depending on whether water is used for processing or cooling.

REQUIREMENTS OF FISH

Fishery requirements (simplified)

The presence of a healthy self-sustaining fish population is usually an indication of a satisfactory environment.

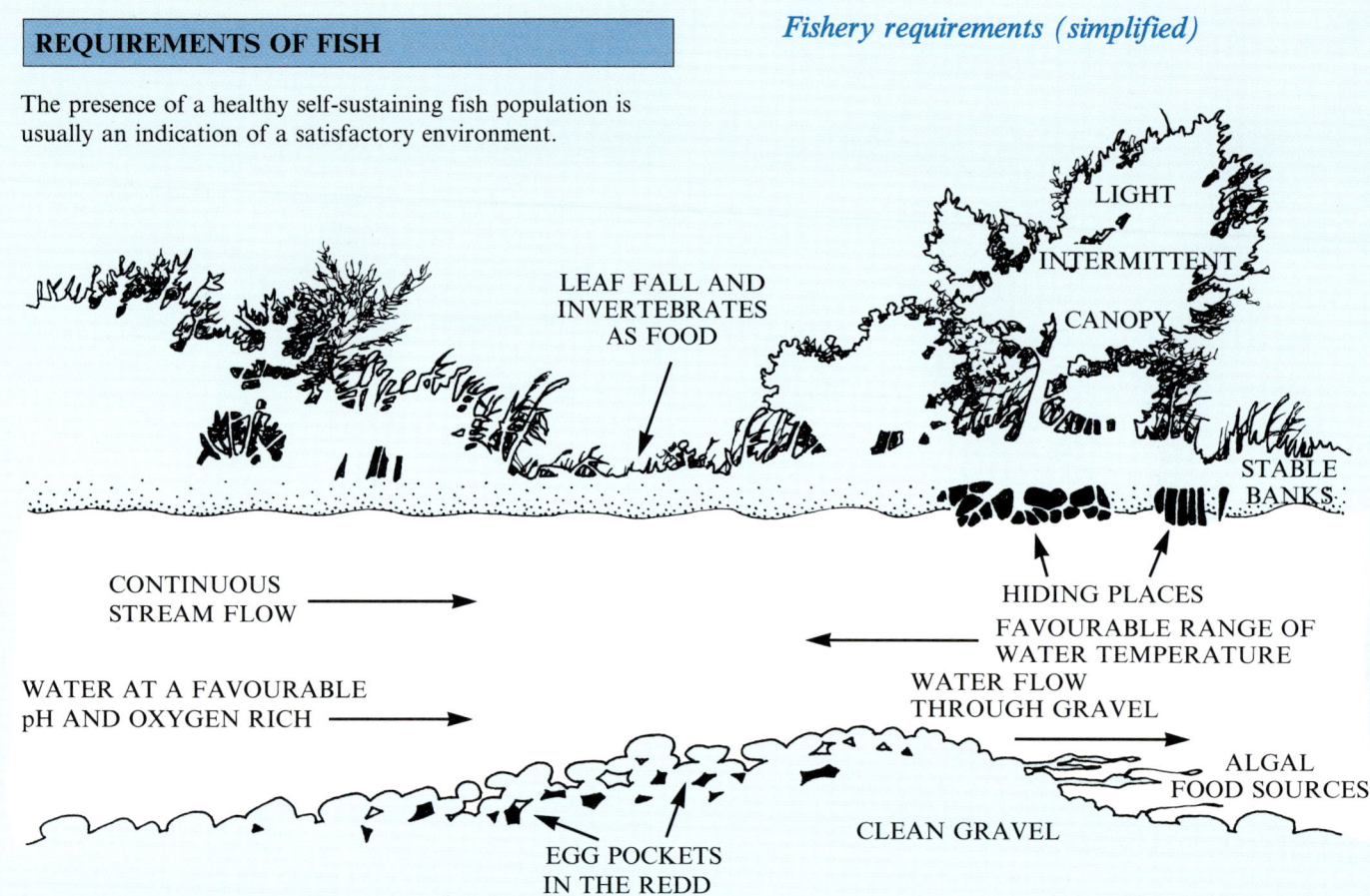

Salmonid fish, which predominate in upland waters, require:

- **clear, cool, relatively silt-free water** for spawning and growth. For normal growth, water temperatures should be in the range 5–15°C and should not exceed 21°C. Some shade is helpful, but not too much;

- **a year-long flow of well-oxygenated water** with a range of depths and velocities to cater for different sizes of fish. Small tributary streams are often important for spawning even though they may not support adult fish;

- **suitable pH** *(ideally 6–9)* and low levels of dissolved aluminium ions. As the pH falls below 6, physiology and growth are increasingly affected both directly, and indirectly through the effect on their food supply of aquatic invertebrates;

- **clean gravel for reproduction.** The survival of incubating eggs and newly hatched alevins depends on the free movement of oxygen-rich water through the 'redds' – patches of gravel in which eggs are deposited. If this flow is blocked by fine sediment, survival rates will fall dramatically. Ideal spawning gravels have less than 15% by weight of sediment material smaller than 1 mm in diameter;

- **reasonably clear water** in the growing season, for sight feeding. Although colour has to be very dark before it hinders sight feeding, this can occur in spate conditions with a heavy load of suspended solids;

- **in-stream cover** – such as deep water, submerged objects, overhanging vegetation, undercut banks – as protection from predators;

- **invertebrates for food;**

- **in migratory species, access to and from the sea and from lakes to feeder streams.** Access is impeded if streams are too shallow, a problem aggravated in small streams by deep gravel deposition which may result in summer flows becoming subterranean. Too high a velocity, high vertical falls or excessive turbidity may also bar upstream migration.

Fish in lowland waters may require different conditions. Some need plants as spawning sites or warm water to allow them to mature. Some are more tolerant of conditions such as lower concentrations of dissolved oxygen or greater turbidity, but this is not always so. Moreover, lowland reaches may contain salmonids or be used by migrating fish. *The rule must be to establish the requirements of the fish communities by consulting the local fishery interests.*

POTENTIAL PROBLEMS

Water yield

Users need adequate quantities of water. Water yields from catchments containing closed canopy conifer forest may be less than from moorland or grassland catchments due to their greater interception loss. This loss increases with forest height and canopy development and is greatest in the wetter, upland areas of Britain. Though assessments of the degree of reduction in a given catchment cannot be exact, research suggests there may be some 1.5–2.0% reduction of potential water yield for every 10% of a catchment under mature forest. Work is in progress to enable a prediction to be made on yield reductions in different areas. Water yields from young forests or felled areas are unlikely to differ significantly from moorland catchments.

The difference in water yield between mature forest and grassland in the lowlands is less than that in the uplands. This is because of the drier and less windy climate, which reduces the size and importance of the interception loss. In the driest lowlands, the difference is believed to be marginal. Evaporation from deciduous broadleaved woodland is less than from conifers or grassland.

The loss in water yield from mature forests is likely to be a problem only in those catchments where the supply is being, or is planned to be, fully exploited. This may include catchments used for the generation of hydro-electricity. Where new planting is proposed in catchments with either hydro-electric schemes or impoundment for water supply or river regulation, early consultation with the water regulatory authority and/or water undertaker will clarify whether there is a problem.

Base flows

Cultivation and drainage of wet soils during establishment can significantly increase summer base flows, an effect that lasts at least until the tree canopy is approaching full closure – 15 or 20 years. The continued growth of a forest could result in decreased summer base flows. The extent to which these two processes balance each other and how this varies over time remains uncertain.

Peak flows

Some users are adversely affected by high peak flows. The ploughing and drainage of a complete catchment has been shown to result in a significant increase in peak flows – perhaps in the order of 20–30%, decreasing to 10% after 10 years – and a decrease in the time to peak of up to one third. These effects occur in moderate rainfall events and decrease with increasing storm size. No increase has been detected in peak flows for very large storm events. However, poor drainage practice could contribute to localised flooding, and altered flow regimes can affect fish migration and angling success.

Sedimentation and turbidity

Erosion resulting from poor management of cultivation, drainage, road building and harvesting can lead to large quantities of sediment entering surface waters. There have been instances where operations have caused unacceptable turbidity levels, seriously disrupting water treatment plants and water supplies. The financial consequences of such incidents can be very great and may even involve the construction of new treatment plants.

High turbidity levels due to inputs of fine sediments such as clay, silt and fine sand can have an adverse impact on the aquatic flora and fauna. Light penetration is reduced, affecting overall productivity, fish feeding and migration. Suspended sediment can also affect fish respiration. When fine sediment settles it can damage spawning areas, by physically blocking or reducing the oxygen supply to fish in their early life stages. Siltation may also blanket plants and modify substrates leading to a decrease in invertebrate diversity.

Large inputs of coarse sediment can also have a serious impact on receiving waters, leading to the destabilisation of stream beds and channels, the shallowing of watercourses, and blockage of pipelines and water intakes to treatment works. Sedimentation may also lead to a long-term reduction in reservoir storage capacity.

Colour, iron and manganese

One particular problem with upland supplies draining from areas with peaty soils, whether forested or not, is the high level of water colour. Colour levels vary greatly from year to year for complex reasons. Climatic factors such as the incidence of drought are believed to be important. The intensity of livestock grazing and moorland drainage work, including drainage prior to new planting, are also thought to play a part.

Increased colour can interfere with water treatment processes and add greatly to treatment costs. It will also reduce light penetration and may thus affect plant growth and productivity. Conversely, it may help to reduce the toxicity of metals such as aluminium by forming non-toxic alumino-organic complexes.

High iron and manganese levels are often associated with the high turbidity levels which can result from the disturbance and erosion of mineral soils following forest operations. This can add to water treatment problems and lead to dirty water supplies, consumer complaints and failure to comply with the EC directive relating to the quality of water intended for human consumption. Iron, in particular, may be precipitated and coat stream beds, leading to harmful effects on the flora and fauna.

Chemicals and fuel oils

Chemicals and fuel oils can taint water supplies and have serious effects on the aquatic environment. Chemicals, and in particular pesticides, can give obnoxious tastes and odours at extremely low concentrations, and markedly increase the cost of water treatment. Water undertakers have a statutory duty to limit the concentration of any individual pesticide in drinking water supplies to less than 0.1 part per billion. Some pesticides can be extremely toxic to fish, aquatic plants and invertebrates, and can build up to damaging levels in birds and other wildlife. The use of pesticides must therefore be kept to a minimum. Very strict attention must be given to preventing spillage or other contamination. Existing controls on the use of pesticides aim to give complete protection. Instructions must be followed meticulously.

Nutrient enrichment

There is concern that the nutrient, and hence the ecological, status of reservoirs and lakes may be significantly changed for a time following the aerial application of phosphate fertilisers in their catchments. Nutrient releases following large scale felling operations may also contribute to ecological changes. The main concern is with upland waters which are naturally nutrient poor and where biological activity is usually phosphorus limited. In extreme cases, phosphorus enrichment can produce excessive algal growths, resulting in a reduction in dissolved oxygen levels and fish deaths. In addition, excess phosphate could require improvements to be made to treatment plants, or result in increased treatment costs.

Heavy rainfall following fertilisation with urea could result in high ammonium concentrations in streamwaters, interfering with water treatment processes and causing an unacceptable taste in drinking water.

A *Establish some broadleaved trees near watercourses.*

B *Maintain about half of stream surface in full sunlight, the rest in dapple shade.*

C *Stop cultivation well short of watercourses.*

D *Do not plough unnecessarily; consider scarifying or mounding.*

FORESTS & WATER GUIDELINES

E *Maintain protective unplanted strips.*
F *Keep branches and tops out of stream and the riparian zone.*
G *Stack timber on dry ground away from watercourses.*
H *Design streamside edges in harmony with the landscape.*

Acidification

The primary cause of acidification is the deposition of acidic sulphur and nitrogen compounds derived from the combustion of fossil fuels. Acidification of fresh waters occurs where the inputs of these pollutants exceed the buffering capacity of the soils and the underlying rocks through which water passes before entering streams, rivers and still-water bodies. The most acidified areas in the UK are in the uplands where catchments with base-poor, slow weathering soils and rocks coincide with high pollutant inputs in the form of large volumes of moderately polluted rainfall. Surface water acidification has been identified as a particular problem in parts of central and south-west Scotland, Cumbria, the Pennines and central and north Wales.

Before the industrial revolution, waters in upland areas were generally unpolluted, with good stocks of salmonids and other fish. Today, where surface water acidification has occurred, fish populations have declined or, in some instances, been lost completely. In many upland areas surface waters are sources of supply for domestic use. Increased acidity and consequent increased solubility of aluminium have implications for public and private water supplies, influencing compliance with water quality standards and potentially increasing the cost of water treatment.

The quantity of sulphur and nitrogen pollutants deposited at a given site is strongly influenced by the nature of the vegetation layer. Forest canopies can significantly increase the capture of some of these pollutants in the atmosphere. This increased capture, often termed scavenging, is a function of the stand structure which creates turbulent air mixing. The effect therefore becomes more important as trees grow and the height of the stand increases. The enhanced capture of mist, which can contain large concentrations of sulphur and nitrogen, is greatest at high altitude because of the increased duration of cloud cover and high wind speeds.

Long-term studies continue in order to determine the magnitude of this scavenging effect and its role in acidification, and also to validate models of deposition. The main pollutant of concern is sulphur due to its established role in surface water acidification. The role of nitrogen remains uncertain since forest growth is characteristically limited by below optimum availability of nitrogen, and drainage water from forests and woodlands generally has very low nitrate concentrations. Consequently nitrogen depositions would not normally be expected to pass through forest ecosystems and result in acidification of water. However nitrate leakage from older forest stands has recently been identified in areas of high nitrogen depositions. Nitrogen depositions and their role in acidification are currently being investigated in a number of research programmes.

Significant nitrate leakage is also known to result from harvesting operations. This is due to increased rates of mineralisation and nitrification in the soil in the absence of uptake by the trees. This pulse may last for between two and four years, depending upon the rate of revegetation. Whilst an increase in nitrate concentrations in soil and stream waters poses only a very small risk of exceeding environmental quality standards, of more concern is the short-term increase in hydrogen ion concentrations which may contribute to acidification and increase aluminium solubility. In acid sensitive catchments, avoiding the clearfelling of large areas in a short space of time reduces the likelihood of such problems.

Acidic depositions

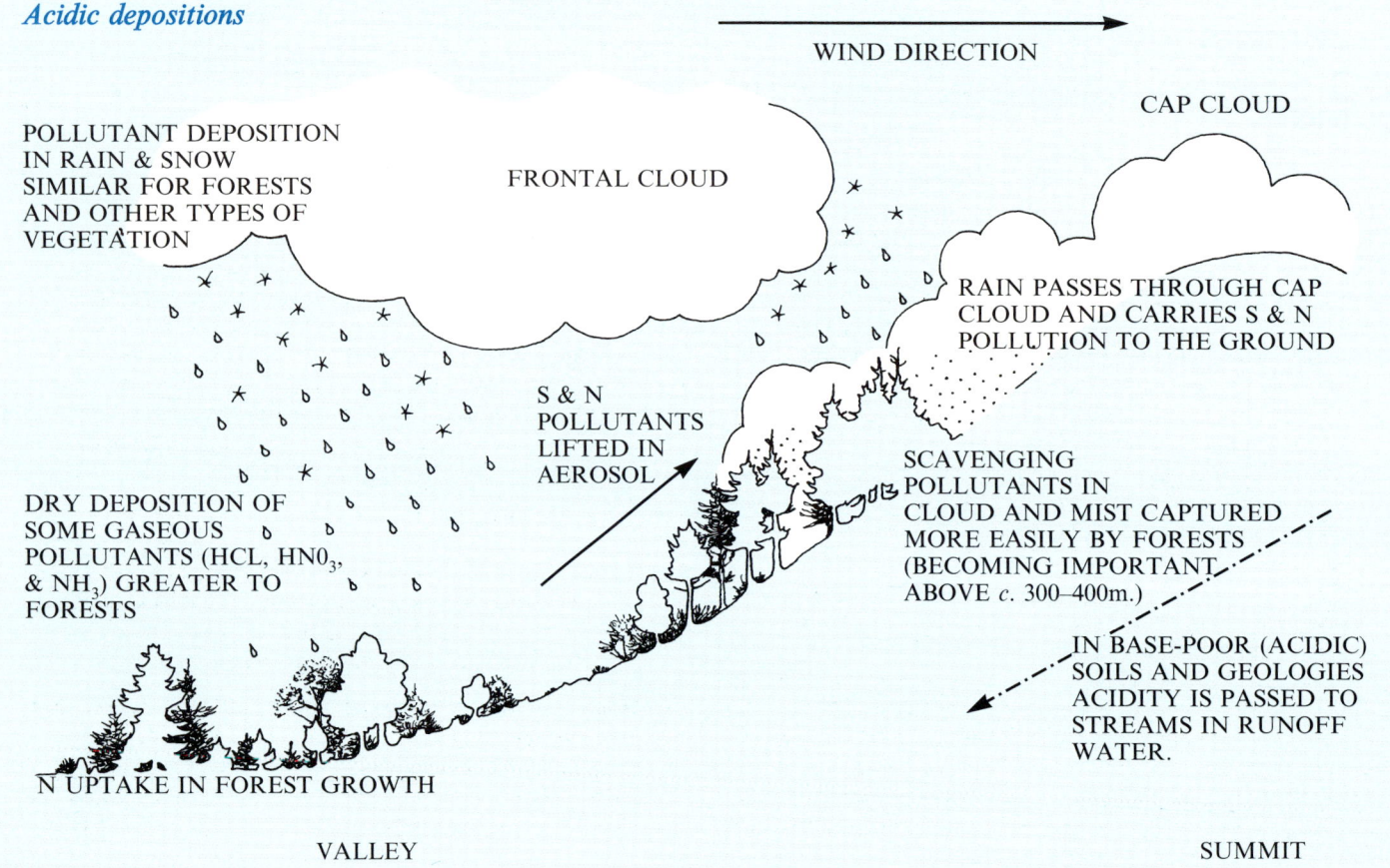

SPECIFIC RECOMMENDATIONS ON FOREST OPERATIONS

CRITICAL LOADS, CATCHMENT ASSESSMENTS AND AMELIORATIVE TREATMENTS

Reduction in the emissions of acid pollutants is the only way of solving the general problem of surface water acidification. In the UK there has been a reduction in emissions, and hence deposition, since the late 1960s and a response to this has already been detected in some surface waters. The European Community (EC) has agreed to further reductions by 2003, but these will not result in the recovery of all acidified waters or remove the risk of continued acidification in all currently susceptible areas. Investigations are under way to assess the effectiveness of agreed emission reductions in reducing pollutant inputs, and to identify the rates and extent of recovery in terrestrial and aquatic ecosystems.

The freshwaters of acid sensitive areas may be at risk from the enhanced capture of acidic pollutants by trees. In order to reach a decision on forestry proposals the Forestry Authority will take into account the effect of scavenging. To this end it is necessary for both the Forestry Authority and applicants to have an understanding of where this effect is likely to be important. The use of critical loads maps provides the most suitable approach. A critical load is defined as: the maximum load of a pollutant which a given ecosystem can tolerate without suffering adverse change. For freshwaters, critical loads can be calculated which, provided they are not exceeded, ensure the maintenance of water chemistry suitable for the protection of populations of fish and other freshwater biota. The Department of the Environment (DoE) has calculated critical loads for freshwaters in the UK. Having compared these with the non-marine inputs of sulphur, the DoE has derived maps which indicate where critical loads for acidity for freshwaters are currently exceeded, and are likely to continue to be exceeded in the year 2005.

The map overleaf is the provisional critical load exceedance for UK freshwaters based on the 1986–88 pollutant (non-marine) sulphur depositions (DoE, 1993). Since depositions are expected to continue to decline over the next 10 years, the use of this map to identify where there may be acidification problems allows for a good safety margin. The map is used here to indicate where additional scavenging by trees could lead to further freshwater acidification. However, because of sampling and scale factors the map is not directly useful for determining the susceptibility of running waters in individual catchments. This requires a catchment based assessment. **Catchment based assessments are likely to be required for new planting proposals within those squares of the map where critical loads are exceeded.**

The need for assessment will be determined by the Forestry Authority taking advice as necessary from the appropriate water regulatory authority. Where an assessment is required these two parties will discuss details with the applicant. In some cases, assessment will be possible on the basis of existing information. Where sufficient information is not already available, assessment is likely to involve the collection of water samples from the main watercourse receiving drainage from the area proposed for new woodland. Samples will be needed to characterise water chemistry at high and low flows, and thus to calculate the catchment's freshwater critical load. The additional pollutant capture by the proposed forest will be estimated and added to the predicted pollutant depositions for the catchment at the time of canopy closure (estimates are currently based on predictions for the year 2005). Where the combined deposition total exceeds the freshwater critical load, approval of a Woodland Grant Scheme is unlikely until there are further reductions in pollutant emissions or unless ameliorative treatments are applied without detriment to the ecosystem. The suitability of any proposed ameliorative treatment will need to be addressed by the Forestry Authority in consultation with the applicant, the water regulatory authority and the appropriate national conservation agency.

Catchment assessments are expected to show that where critical load exceedance is currently greatest (i.e. by more than 0.5 keq. H^+ ha^{-1} yr^{-1} as in the black and grey squares) and a major proportion of the proposed planting is at higher altitude, the scavenging effect will result in critical load exceedance at canopy closure. Because sulphur depositions have declined and are expected to continue to decline, pollutant scavenging may not be a significant problem in areas where critical load exceedances are currently small (the light blue and dark blue squares of the map).

GROUND PREPARATION

Cultivate only those parts of a site where it is necessary. Plan all cultivation and drainage meticulously. The following measures will minimise any effects. Further useful advice is given in Forestry Commission Research Information Note 196 *Forest drainage*.

- **For new planting in the uplands, the use of scarifiers for dry soils and the use of continuous-acting mounders – preferably fitted with a moling or ripping attachment – for wet soils, is recommended for all soils except the wettest peaty gleys and peats.** On peaty soils, spaced furrow ploughing should be as shallow as possible (e.g. 30 cm), aiming to expose mineral soil as little as possible. Most mineral soils are more erodible than peats, and they can be a source of toxic aluminium entering watercourses in acid sensitive areas. Loose sandy or loamy soils are more erodible than compact soils or clays. If exposing mineral soil is unavoidable a plough without a tine is preferred. In the lowlands, cultivation is unlikely to be necessary except, perhaps, for weed control.

- **For establishing new native woodlands** minimal cultivation is especially important. If cultivation is considered essential for establishment, mounding alone, without ripping or moling, should always be sufficient, and ploughing is to be avoided. Very wet soils should either be left unplanted or mounded and planted with a species appropriate for such conditions. Drains should be used only as a remedial measure if there is a need to control discharges from pre-existing natural or man-made channels.

- **When restocking,** cultivation is unlikely to be required except on iron-pan soils or for weed control. Scarifiers are appropriate for better drained soils, whereas mounders are suitable for wetter sites. On very wet as well as weedy sites, excavator mounding can be used to provide intensive drainage as well as good planting positions.

- **Where excavator mounding** is used to cultivate restock sites and exposes mineral soil, prevent spoil ditches from carrying large volumes of water by ensuring that individual ditches are no longer than 30 m. On long slopes the ditches of consecutive 30 m bands should be staggered across the

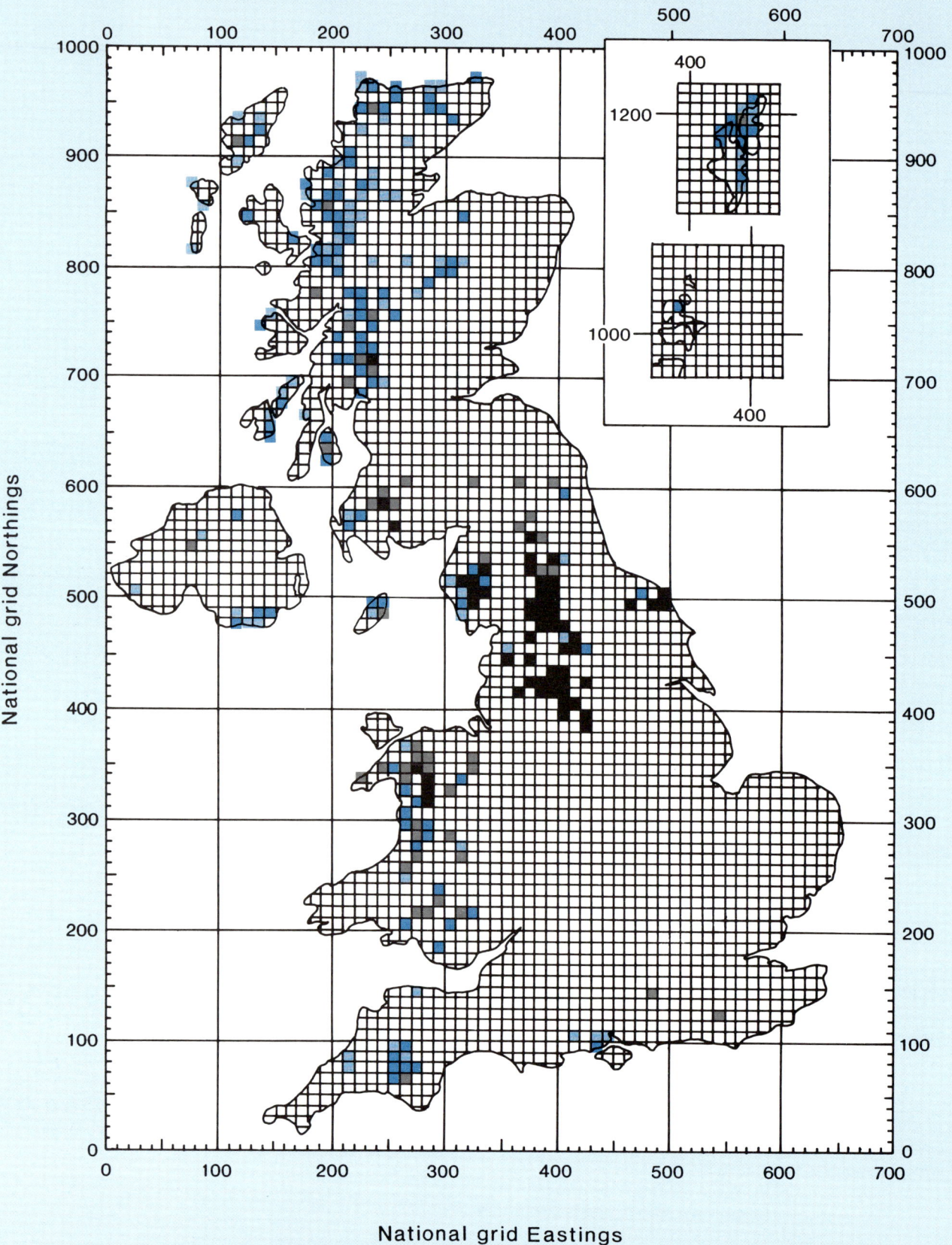

Provisional critical loads exceedance for UK freshwaters (Henriksen model), March 1993. Exceedance (keq H^+ ha^{-1} year^{-1}): 0.0–0.2, 0.2–0.5, 0.5–1.0, >1.0. Not exceeded white.

Forest drainage

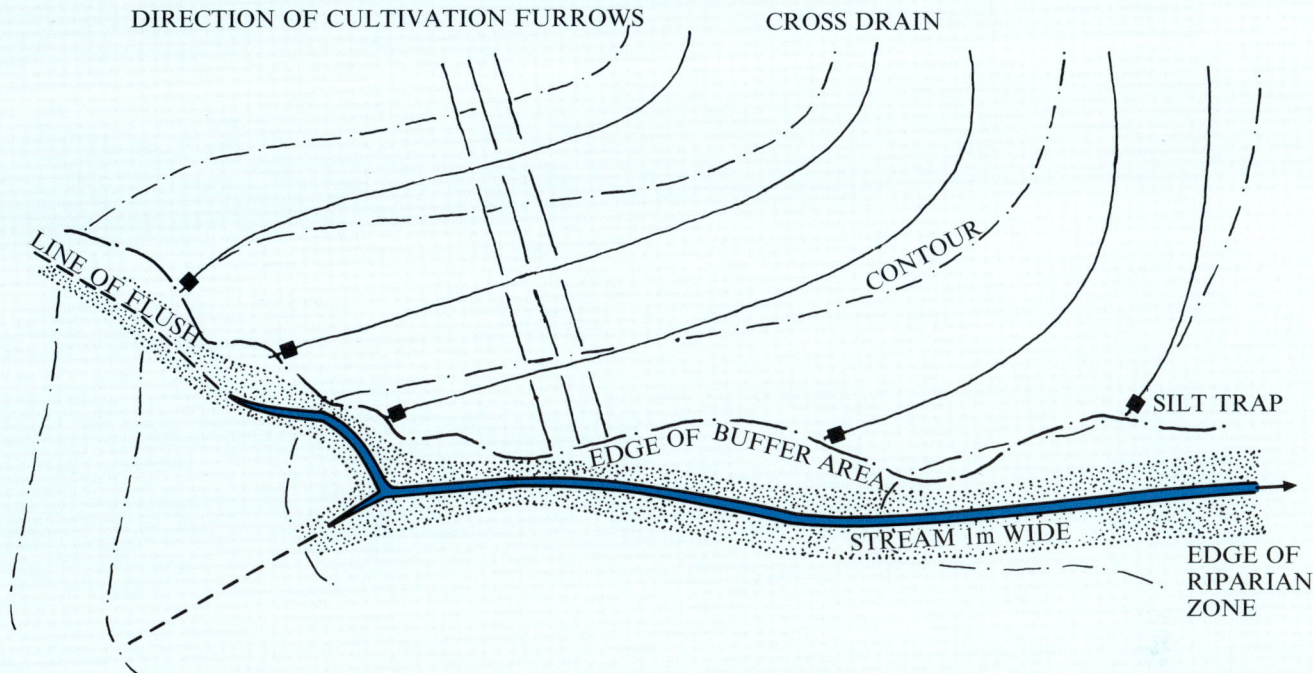

slope and an unditched (but mounded) buffer strip left between the bands.

- **Provide cross drains** at a spacing that will control runoff in cultivation channels, e.g. 40 m on slopes less than about 5% (3°). Provide cut-off drains so that cultivation channels do not carry water from large areas lying above.

- **When ploughing and cross-draining** on slopes over about 9% (5°), lift the plough above each pre-marked drain line to provide a clean turn-out and a local sediment trap. The drain line should include a 3–5 m wide strip of vegetation between the ends of the furrows and the drain itself. On complex terrain this may not be important or, indeed, practical. This strip can be planted, i.e. on turfs. When ripping or moling on wet soils, the cultivation channels should intersect the drain.

- **Align drains up-valley** so as to maintain an even gradient throughout their length (see Figure). Drain gradients should not exceed 3.5% (2°) and should be less on easily erodible soils.

- **Keep drain ends back** from the riparian zone and especially from steep gully sides bordering that zone or the stream. (See BUFFER AREAS, below.)

- **Install drains immediately after cultivation** especially where large volumes of runoff could reach a new forest road. Roadside drains will often exceed the permissible gradient for cross drains and cannot withstand large volumes of water. Try to avoid any water other than that from the drainage of roads entering roadside drains.

- **Construct silt traps** at the ends of drains in areas of high or moderate erosion risk, ensuring that there is machine access for periodic emptying. Spoil should not be dumped on the floodplain or on ecologically valuable wetlands (see Figure).

- **Never divert natural watercourses,** however small or ephemeral, into cultivation channels or drains.

- **Avoid diverting drains to adjacent catchments** so as not to overload the recipient stream. The risk is greatest in high rainfall areas. Even where there is a requirement to avoid disturbance to water courses, it may be safer to carry out any necessary drainage, with particular care, within the catchment than to route drains to an adjacent catchment.

- **Organise drains maintenance and silt trap cleaning to avoid the spawning season or the period when salmonid eggs and alevins are living in the gravel.** The sensitive period varies a little from place to place, but October to May inclusive should cover the main spawning and incubation periods in the uplands. In lowland rivers, consult the local fishery interest.

BUFFER AREAS (see figures on this page and pages 4 & 8/9)

A buffer area is an area of uncultivated land which protects the riparian zone and watercourse from discharges off the adjacent land, particularly from drains. The buffer area will generally encompass the riparian zone and might extend beyond it. The buffer area filters out sediment from the water passing through it, and may also reduce the concentration of nutrients. The effectiveness of the buffer area is increased if it is wide and gently sloping and if the vegetation is vigorous. For advice see *Forest nature conservation guidelines* and *Forest landscape design guidelines*.

- **Discharge from a drain** should, as far as possible, be on flat ground so that the water can fan out rather than be allowed to emerge in a concentrated flow. Do not end drains in natural channels, ephemeral streams or old ditches.

- **The width** of the buffer area should be determined by the risk of sediment movement, which, in turn, will depend on the type of cultivation, erodibility of the soil, gradient and catchment area of the drains. An indication of average widths is as follows:

  - for a headwater stream with a channel up to 1 metre wide, a buffer at least 5 metres wide on either side;
  - for a stream with a channel 1 to 2 metres wide, a buffer about 10 metres wide on either side;
  - for a stream with a channel over 2 metres wide, a buffer about 20 metres wide on either side.

  For very erodible soils these widths should be doubled. For environmental reasons the width of the buffer must not be constant and should be widest at the point of discharge from drains.

  How much of the buffer area should bear trees will be determined by ecological, aesthetic and practical considerations. Suffice to say here that ground vegetation has a crucial role in trapping sediment and clearly must not be allowed to be shaded out by the mature canopy.

MANAGING RIPARIAN VEGETATION

Because the structure and composition of riparian vegetation greatly influence the adjoining and downstream aquatic environments, the location and species composition of riparian woodland are of great consequence. Riparian zones should be managed in an integrated way so that the aquatic environment and water quality are protected or enhanced and the value of distinctive riparian habitats is maintained or enhanced. Full details of riparian zone management are given in *Forest nature conservation guidelines*.

Riparian management should:

- **apply to all features which are characteristic of riparian land** including headwater source areas, gully banks, river terraces, flood plains, swamps and carrs.

- **maintain open or partially wooded conditions** such that bankside vegetation thrives, thereby minimising bank erosion and opening up the water to sunlight. The exclusion of stock will generally be helpful in reducing both the erosion of stream banks and the overgrazing of streamside vegetation.

- **maintain about half the length of a stream open to sunlight,** the remainder being under dappled shade from trees and shrubs. Periodic management may be needed to maintain a sensible balance between light and shade. Trees should not be planted on, or allowed to regenerate into, ecologically rich open ground habitat. Remember that open ground south of a stream is most valuable.

Choice of species

Heavily shading trees such as oak, beech and many conifers should be used sparingly. Densely shading trees should generally be interspersed with lighter foliaged trees such as birch, willow, rowan, ash, hazel, aspen and bird cherry. Litter from these species decomposes more rapidly than litter from oak, beech and most conifers, and is therefore more valuable to aquatic invertebrates, especially in small streams. Alder should be used cautiously; although its roots can help to bind banksides and can contribute nutrients by fixing nitrogen, it casts a heavy shade and may contribute to acidification.

Where the riparian zone has been planted with inappropriate species or too densely, the undesirable trees should be removed when a management opportunity arises. Trees growing on particularly steep or unstable sites need careful management in order not to precipitate erosion or slope failure; it may be sensible to leave them standing. Natural regeneration, particularly of heavily shading trees, may need attention though in general natural regeneration of site native species should be favoured.

Deer management

A well managed riparian zone can be very valuable for deer management, providing opportunities for linked deer glades and for establishing browse species.

Wildlife management

The richness of wildlife within a woodland or forest often depends on both the quantity, quality and distribution of open ground and the transition zones between it and the trees. Together these provide that structural diversity which is attractive to woodland fauna and to the plants which flourish in semi-woodland conditions. See *Forest nature conservation guidelines* for further information.

Landscape design

Vary the width of the riparian zone. Introduce variations in the density and species composition, not only in the zone but also in the transition to dense forest, so that the whole composition is in harmony with the landscape. Use slower growing conifers such as Scots pine or larch on streamside knolls, especially on the north side of streams; these can be valuable for wildlife if retained to an advanced age. There may be a requirement for a limited amount of planting almost to the water's edge to obtain satisfactory visual linkage across a stream. See *Forest landscape design guidelines* for further information.

Imaginative management of the riparian zone will make a vital contribution to multi-purpose forestry.

Roadside drains

ROAD CONSTRUCTION AND MAINTENANCE

Forest road work requires meticulous planning and execution if damage to streams is to be avoided.

The following measures help to achieve this.

- **Build roads well clear of riparian zones** wherever possible.

- **Avoid metalliferous or sulphide-rich material** for road construction near watercourses. Acid metal-rich drainage water can be extremely toxic to aquatic organisms.

- **Roadside drains** should not intercept large volumes of water from ground above. Any watercourse, however small, that is intercepted by a road should be culverted or bridged at that point. If additional culverts are needed to discharge water from the roadside drain, they should be of a size sufficient to avoid overloading, blocking or washout. Roadside drains, likely to carry high sediment loads, must not be allowed to discharge directly into streams, but must discharge to a buffer area of adequate width (see BUFFER AREAS). Drains on the upper side of the road may need culverts to the lower side a short distance before stream crossings so as to prevent direct discharge. (See Figure).

- **Where appreciable sediment** movement is unavoidable, silt traps should be constructed and provision made for maintaining them. Cleaning should be done in dry weather during May to September, to avoid the salmonid spawning period.
- **Culverts** should be protected and well bedded to avoid settlement.
- **Erosion should be prevented at both ends of culverts** by reinforcing the substrate and reducing flow velocities.
- **Down flumes** or other suitable structures should be provided at the outflows of hanging culverts, with rock or concrete aprons as stilling areas.
- **Culverts in fish-bearing streams** must not be a barrier to migrating fish. The aim should be to produce a culvert floor at the same gradient and level, and carrying similar bed material and flow, as the original stream. A bridge may be necessary if these conditions cannot reasonably be met.
- **Avoid erosion** of roadside embankments and cuttings by using intercepting trenches or terracing. Keep embankments and cuttings at no greater slope than the natural angle of repose to encourage revegetation, otherwise added stabilisation will be needed. Consider revegetating exposed soil artificially if natural re-establishment is slow. Such measures are especially important near culverts. Similar considerations apply to borrow pits.

During construction

- **It is of over-riding importance** where fisheries are significant, that any in-stream work must be carried out during periods when there is the least likelihood of damage to fish. May to September is a favourable period for such work if damage to eggs and alevins is to be avoided, but other local factors may have a bearing on the decision. Seek advice from the local fishery interest.
- **Consult** the water regulatory authority, other riparian owners, tenants and fishery interests before extracting gravel from a watercourse. On many rivers consent is required; in some rivers gravel extraction is illegal.
- **Wherever practicable** avoid construction near streams in periods of wet weather.
- **Keep cement or raw concrete** out of watercourses when constructing bridges or crossings. Cement can be lethal to fish and other aquatic life.

During maintenance

- **Keep roadside drains** and culverts clear of debris to avoid blockage and wash-out; avoid unnecessary disturbance of the vegetation along the roadside drain.
- **Inspect silt traps regularly** and clear out as needed. (See *silt traps* in GROUND PREPARATION.)
- **Carry out maintenance** of drains and silt traps in dry weather, avoiding fish spawning periods.
- **Maintain an adequate cross-camber** and prevent the formation of wheel ruts; they will erode, and road surface material will be carried into the drains.
- **Maintain vegetation** on embankments.
- **Observe the** *Provisional code of practice for the use of pesticides in forestry* if herbicides are to be used to control vegetation in drains and on roads.
- **Inspect roads** and associated drainage during, or soon after, intense rainstorms.
- **Inspect roads, drains and silt traps** before and after any intensive use of forest roads such as harvesting or rallying, and take remedial action quickly.

HARVESTING

Good planning and management, including on-site discipline and attention to detail, can do much to minimise any adverse effects of harvesting operations on streams. Make sure that felling and extraction teams and supervisors fully understand and observe any special working instructions.

Felling operations can disturb the soil and remove part or all of the intercepting canopy, allowing more precipitation to reach the ground. Exposure and disturbance of the soil are determined by the scale of operation and the equipment used. Small scale operations such as thinning or group felling are likely to involve smaller mechanical equipment and cause less impact. Large scale clearfelling poses the greatest risks. Potential dangers to be avoided are: significant increases in sedimentation, damage to spawning grounds, blockage of streams, prevention of fish movement and ingress of excessive amounts of bark or woody debris which can smother the stream bed.

- **Liaise with the local water regulatory authority and water undertaker** at the early planning stage when harvesting in water supply catchments. In acid sensitive and water supply catchments consider phased felling. In large catchments it is unlikely that felled areas will be more than a small fraction of the whole.
- **Locate any private or public water supplies** and give them the necessary protection.
- **Choose dry sites for stacking timber,** well away from watercourses. Do not block roadside drains.
- **Plan felling and extraction** to minimise the number of stream and drain crossings. Do not plan a felling coupe which includes both sides of a watercourse unless there is road access to each side. Where crossing a stream or drain is unavoidable install a log bridge or piped crossing. Even small runnels which may be dry before felling may flow again during operations; consider them for piped crossings.
- **Choose the best machine combination** for the ground conditions including appropriate traction or flotation aids. Avoid ground skidding on soft soils. Avoid long ground-extraction routes on steep slopes, especially in high rainfall areas.
- **Cable-crane extraction,** particularly of shortwood, causes much less soil disturbance than skidding or forwarding; consider it for sensitive catchments.
- **On soft soils** provide and maintain an adequate supporting brash mat for the principal vehicle routes. Brash may have to be transported from where it is plentiful. Where forest roads have to be used for long distances by forwarders and the like, use brash thatching to prevent damage to the road.

- **On sensitive sites** try to work during spells of good weather.

- **Felling in the riparian zone** will be an infrequent event once the recommended vegetation has become established; where it is needed, fell trees away from the stream. Keep streams free from branches and tops as far as practicable. An occasional large log in a stream may be advantageous in creating a pool, provided it is reasonably stable and does not cause erosion by diverting the stream.

- **When extraction tracks** have been created on slopes, prevent water running down any wheel ruts by digging offlets at intervals. Make this also the standard practice if there is any appreciable break in operations, or during operations if there is a risk of erosion.

- **Make sure that** haulage roads, drains and culverts are adequate and in good repair before work starts. Never let extraction and haulage machines destroy a sub-standard road in the hope that it can be repaired cheaply afterwards – erosion can be very serious, and either expensive or impossible to put right.

- **Modify operating procedures immediately** if erosion is occurring and construct silt traps if necessary.

- **Avoid fuel spillages.** Carry out machine refuelling and maintenance well away from watercourses. Prepare and install a contingency plan.

- **Provide adequate protection** against damage by machinery and vehicles for buried water conduits and pipelines.

- **Prompt restocking** after harvesting will help to stabilise the site.

PESTICIDES

Forest managers should refer to the *Provisional code of practice for the use of pesticides in forestry* for detailed guidance.

Before using pesticides which might get into water it is recommended that there should be prior liaison with the water regulatory authority. Consultation with the water regulatory authority is **legally required** before the use of any pesticide in or near water and before the aerial application of pesticides on land adjacent to water under the Control of Pesticide Regulations (1986) and Control of Substances Hazardous to Health Regulations (1988). Consultation should take place at an early planning stage wherever possible, and legally must occur not less than 72 hours before an application begins. The water undertaker should also be notified well in advance of any pesticide application to land in close proximity to watercourses or water bodies used for public water supply.

The **consent** of the water regulatory authority and undertaker is **required** if an aerial application of herbicide is intended for the control of aquatic weeds or weeds on the banks of watercourses or lakes. The MAFF *Guidelines for the use of herbicides on weeds in or near watercourses and lakes* requires that only those products which carry specific recommendations for aquatic weeds should be used and only with the prior agreement of the water regulatory authority.

Key points for action are:

- **Consult with, and obtain consent if necessary from, the water regulatory authority.**

- **Prepare a contingency plan** to deal with accidental spillage.

- **Store chemicals securely** in locations away from watercourses. Guard against accidental spillage.

- **Read container labels** and follow the instructions meticulously.

- **Try to avoid overspraying drains.**

- **Do not spray or apply granules within 10 metres of watercourses** and within 20 metres of lakes and reservoirs.

- **Spray only when wind conditions** are appropriate and with the correct droplet size to minimise drift.

- **Do not apply chemicals** if heavy rain is expected or if there is a risk of wash-off because the ground is frozen, snow-covered, or baked dry.

- **Do not wash out sprayers, containers or the like** near any watercourse, however small.

- **Do not puncture or bury empty containers.**

- **Seek advice** from the appropriate water regulatory authority about the safe disposal of unwanted pesticides and containers.

- **Never store or soak planting stock which has been treated with an insecticide in a watercourse** prior to planting. Certain insecticides, e.g. Permethrin, are extremely toxic to fish and other aquatic life.

FERTILISERS

Key action points when applying fertilisers are:

- **Contact the water regulatory authority and water undertaker when planning aerial fertiliser applications in catchments with standing waters.** Establish the sensitivity of lakes, reservoirs or other water bodies and agree the plans for fertiliser application.

- **In catchments of sensitive water bodies** consider applying fertiliser by hand or ground machine, or phasing aerial treatments over several years.

- **Do not apply fertilisers** in very wet weather or if heavy rain is forecast. Do not apply if the ground is frozen, snow-covered, or baked dry, when the risk of wash-off is at its greatest.

- **Avoid riparian and aquatic zones.**

- **Store fertilisers** away from all watercourses.

- **Prepare a contingency plan in case of spillage.** Contact the water regulatory authority in the event of a spillage and take immediate action to gather up the spill and prevent the contamination of watercourses.

The possibility of using sewage sludge as a forest fertiliser has received increasing attention in recent years. While sewage sludge can be a relatively cheap and useful form of fertiliser and there are environmental advantages to this method of disposal, its use is restricted to certain types of site due to the risk of contaminating surface run-off and groundwater resources. Forest managers should refer to *A manual of good practice for the use of sewage sludge in forestry* for detailed guidance.

Key action points when applying sewage sludge are:

- **Contact the water regulatory authority and water undertaker** when planning sewage sludge applications.

- **Apply sludge only** to sites which satisfy the criteria detailed in the manual of good practice. Do not apply when the water-table is near the surface or when the soil is saturated.

- **Do not apply in wet weather** or if heavy rain is forecast.

- **Do not apply to the riparian or aquatic zones** nor within 50 m of a spring, well or borehole.

- **Prepare a contingency plan in case of spillage.** Contact the water regulatory authority in the event of a spillage and take immediate action to gather up the spill and prevent the contamination of watercourses.

STORAGE AND HANDLING OF CHEMICALS AND FUEL OILS

Plan for careful use of fuels, oils, and chemicals, with proper attention to storage and refuelling areas. These substances can have a serious effect on aquatic life and can taint water supplies and must be kept out of drains and watercourses. Fuels should be stored in bunded or double-skinned tanks and refuelling carried out using a transfer hose. Tanks should be locked when unattended. Advice on secure storage arrangements can be obtained from the water regulatory authority.

CONTINGENCY PLAN IN CASE OF CHEMICAL OR FUEL OIL SPILLAGE

A contingency plan should be drawn up to deal with accidental spillages. The plan should include relevant telephone numbers (water regulatory authority, downstream landowners, water users and water undertaker; see Appendix I) and record the availability of equipment to carry out remedial work in advance of the arrival of the water regulatory authority. For example booms and absorbent sheets and pillows should be available to contain and absorb spillages and to prevent them entering nearby watercourses. Machine operators should carry a small supply of absorbent sheets and pillows in their cab, and there should be central stocks of materials and equipment packed and ready for use in emergency trailers. **For further information read the Forestry Authority Technical Development Branch Report No 7/93; Tel: 0387 86264 for copies.**

PONDS FOR FIRE-FIGHTING OR WILDLIFE

Fire-fighting plans may require the construction of artificial ponds or water-holes to ensure a ready supply of water; ponds may also be constructed for conservation purposes. These should be constructed by diverting water into an excavated pond to one side of a stream channel, not by damming the stream itself. The stream is thus kept open, fish are not impeded and the pond does not silt up so quickly. The provision of an island will enhance the wildlife value of a pond. Construction of dams in streams **requires a licence** (from the NRA in England & Wales – in Scotland the appropriate RPA should be contacted) and there is a **legal requirement** not to impede the passage of migratory fish. The construction of ponds and any maintenance work should be carried out in dry weather during May to September to avoid salmonid spawning and incubation periods.

WORKING CHECKLIST

This checklist may be freely reproduced except for sale or advertising purposes.

All operations

- Consult the water regulatory authority and, where appropriate, the water undertaker at the planning stage, to establish the sensitivity of the catchment and any legal requirements.
- Refuel and maintain machinery well away from watercourses; guard against spillage.
- Prepare a contingency plan in case of accidental spillage of fuels, oils or chemicals.
- Make sure that fuels, oils and chemicals are stored safely and away from watercourses.
- Ensure that working instructions are understood and observed.

Ground preparation

- Ensure that all operators know the contingency plan.
- Scarifying or mounding is recommended on all but the wettest peaty gleys and peats.
- Do not plough deeper than is necessary; a maximum of 30 cm is recommended.
- Ensure that individual ditches created on restock sites are no longer than 30 m.
- Stop plough furrows and drain ends well short of watercourses.
- Provide cross drains at a spacing that will control runoff from cultivation channels.
- Align drains so that the gradient does not exceed 3.5% (2°).
- Align drains up-valley to maintain an even gradient.
- Leave drain-side buffer areas at the ends of plough furrows on slopes over 9% (5°).
- Prepare silt traps or pools where there is a high risk of erosion.
- Confine drains maintenance to the summer to avoid fish spawning periods or alevins living in the gravel.
- Do not divert natural watercourses into drains.

Riparian management

- Maintain buffer areas of varying width on each bank of streams and rivers.
- Maintain about half the length of a stream open to sunlight, with the rest under dappled shade from appropriate trees and shrubs.
- Design the transition from buffer area to forest in harmony with the landscape.

Road construction and maintenance

- Ensure that all operators know the contingency plan.
- Build roads well clear of riparian zones wherever possible.
- Avoid construction near streams in wet weather.
- Avoid metalliferous or sulphide-rich road construction material.
- Plan so that roadside drains do not intercept large volumes of water from ground above.
- Plan so that roadside drains do not discharge directly into watercourses, but rather through a buffer area of adequate width.
- Make culverts big enough and install anti-erosion measures.
- Install culverts which will not obstruct the passage of fish.
- Do any essential in-stream work when damage to fisheries is least likely, i.e. in the period May to September.
- Consult the water regulatory authority and other interests before extracting gravel from a water course.
- Keep cement and raw concrete out of watercourses.
- Keep drains, culverts and silt traps clear of debris and carry out maintenance only in dry weather.
- Ensure that road cambers are adequate and carriageways rut free.
- Maintain vegetation on embankments.
- Observe the *Provisional code of practice for the use of pesticides in forestry* when controlling vegetation in drains with herbicides.
- Inspect roads, drains and silt traps for damage after intense storms and also before and after any intensive use such as extraction or rallying.

Harvesting

- Liaise with the water regulatory authority and the water undertaker at the planning stage.
- Ensure that all operators know the contingency plan.
- Locate private and public water supplies and give them the necessary protection.
- Consider phased felling or reducing the scale of operations.
- Fell and extract in sensitive areas during dry weather.
- Protect underground culverts and pipelines.
- Avoid skidding on soft soils.
- Choose dry sites for stacking timber well away from watercourses; do not block roadside drains.
- Plan extraction to minimise stream crossings.
- Use pipes or a log bridge where extraction routes must cross watercourses.
- Avoid long ground extraction routes on steep ground, especially in high rainfall areas.
- Do not let machines work in streams.
- Fell trees away from streams.
- Keep branches and tops out of streams.
- Use brash mats wherever necessary to protect the soil.

Pesticide, fertiliser and sewage sludge application

- Contact the water regulatory authority before work starts.
- Ensure that all operators know the contingency plan.
- Store chemicals securely, well away from watercourses.
- Read container labels and follow the instructions meticulously.
- In and around water supply catchments, use only insecticides and herbicides approved for use in these locations.
- Pay attention to short-term weather forecasts and do not apply when heavy rainfall is expected.
- Do not spray over watercourses.
- Do not apply within 10 m of streams or 20 m of reservoirs.
- Use correct spray dosage rates.
- Do not wash out equipment near drains or watercourses.
- Do not puncture or bury empty containers.
- Seek advice from the water regulatory authority on the safe disposal of empty containers and unwanted pesticides.
- Do not soak insecticide-treated plants in drains or streams.

© *Crown copyright 1993*

APPENDIX I: SOURCES OF ADVICE

In England and Wales forest managers should obtain advice on matters relating to the water catchments in their areas from the appropriate regional office of the National Rivers Authority, and also from the regional water company in the case of water supply catchments. In Scotland such advice is obtainable from the appropriate river purification authority, the water department of regional and Islands councils, the Central Scotland Water Development Board, Scottish Power and Scottish Hydro-Electric. Appendix II gives an outline of the roles of these bodies. **The local conservancy office is the contact point for advice from the Forestry Authority.**

Forest managers are strongly advised to contact the local office of their water regulatory authority and/or water undertaking to obtain the emergency telephone numbers applicable to their area of operations, and then to ensure that site supervisors know these numbers.

WATER REGULATORY AUTHORITIES

Scotland

Orkney Islands Council
Environmental Health
  Department
Council Offices
School Place
Kirkwall
Orkney
KW15 1NY

Tel: (0856) 873535

Shetland Islands Council
Department of
  Environmental Services
Grantfield
Lerwick
Shetland
ZE1 0NT

Tel: (0595) 2024

Western Isles Islands Council
Environmental Health
  Department
Sandwick Road
Stornoway
Isle of Lewis
PA87 2BW

Tel: (0851) 703773

**Clyde River Purification
  Board**
Rivers House
Murray Road
East Kilbride
Glasgow
G75 0LA

Tel: 58 (035 52) 38181–6

**Forth River Purification
  Board**
Clearwater House
Heriot-Watt Research Park
Avenue North
Riccarton
Edinburgh
EH14 4AP

Tel: (031) 449 7296

**Highland River Purification
  Board**
Strathpeffer Road
Dingwall
IV15 9QY

Tel: (0349) 62021

**North East River
  Purification Board**
Greyhope House
Greyhope Road
Torry
Aberdeen
AB1 3RD

Tel: (0224) 248338

**Solway River Purification
  Board**
Rivers House
Irongray Road
Dumfries
DG2 0JE

Tel: (0387) 720502

**Tay River Purification
  Board**
1 South Street
Perth
PH2 8NJ

Tel: (0738) 27989

**Tweed River Purification
  Board**
Burnbrae
Mossilee
Galashiels
TD1 1NF

Tel: (0896) 4797

NATIONAL RIVERS AUTHORITY

England and Wales

Head Office
Rivers House
Waterside Drive
Aztec West
Bristol
BS12 4UD

Tel: (0454) 624 400
Fax: (0454) 624 409

London Office
30–34 Albert Embankment
London
SE1 7TL

Tel: (071) 820 0101
Fax: (071) 820 1603

Anglian Region
Kingfisher House
Goldhay Way
Orton Goldhay
Peterborough
PE2 0ZR

Tel: (0733) 371 811
Fax: (0733) 231 840

Northumbria & Yorkshire Region
21 Park Square South
Leeds
LS1 2QG

Tel: (0532) 440 191
Fax: (0532) 461 889

North West Region
Richard Fairclough House
Knutsford Road
Warrington
WA4 1HG

Tel: (0925) 53999
Fax: (0925) 415 961

Severn Trent Region
Sapphire East
550 Streetbrook Road
Solihull
B91 1QT

Tel: (021) 711 2324
Fax: (021) 722 5824

Southern Region
Guildbourne House
Chatsworth Road
Worthing
West Sussex
BN11 1LD

Tel: (0903) 820 692
Fax: (0903) 821 832

South West and Wessex Region
Manley House
Kestrel Way
Exeter
EX2 7LQ

Tel: (0392) 444 000
Fax: (0392) 442 005

Thames Region
Kings Meadow House
Kings Meadow Road
Reading
RG1 8DQ

Tel: (0734) 535 000
Fax: (0734) 500 388

Welsh Region
Rivers House
Plas-yr-Afon
St Mellons Business Park
St Mellons
Cardiff
CF3 0EG

Tel: (0222) 770 088
Fax: (0222) 798 555

WATER UNDERTAKERS

Scotland

Orkney Islands Council
Department of Engineering
 and Technical Services
Council Offices
School Place
Kirkwall
Orkney
KW15 1NY

Tel: (0856) 873535

Shetland Islands Council
Department of
 Environmental Services
Grantfield
Lerwick
Shetland
ZE1 0NT

Tel: (0595) 2024

**Western Isles Islands
 Council**
Department of Engineering
 Services
Sandwick Road
Stornoway
Isle of Lewis
PA87 2BW

Tel: (0851) 703773

Borders Regional Council
Department of Water and
 Drainage Services
West Grove
Waverley Road
Melrose
TD6 9SJ

Tel: (089682) 2056

Central Regional Council
Department of Water and
 Drainage
Woodlands
St Ninian's Road
Stirling
FK8 2HB

Tel: (0786) 64213

**Central Scotland Water
 Development Board**
Balmore
Torrance
by Glasgow
G64 4AJ

Tel: (0360) 20511

**Dumfries and Galloway
 Regional Council**
Water and Sewage
 Department
English Street
Dumfries
DG1 2DD

Tel: (0387) 61234

Fife Regional Council
Department of Engineering
Fife House
North Street
Glenrothes
KY7 5LT

Tel: (0592) 754411

Grampian Regional Council
Department of Water
 Services
Woodhill House
Westburn Road
Aberdeen
AB9 2LU

Tel: (0224) 682222

Highland Regional Council
Department of Water and
 Sewerage
Regional Buildings
Glenurquhart Road
Inverness
IV3 5NX

Tel: (0463) 702000

Lothian Regional Council
Department of Water and
 Drainage
6 Cockburn Street
Edinburgh
EH1 1NZ

Tel: (031) 229 9292

**Strathclyde Regional
 Council**
Water Department
419 Balmore Road
Glasgow
G22 6NU

Tel: (041) 336 5333

Tayside Regional Council
Water Services Department
Bullion House
Invergowrie
Dundee
DD2 5BB

Tel: (0382) 562581

England and Wales

Anglian Water plc
Ambury Road
Huntingdon
Cambs
PE18 6NZ

Tel: (0480) 433433

Northumbrian Water plc
Abbey Road
Pity Me
Durham
DH1 5EZ

Tel: (091) 384 4222

North West Water plc
Dawson House
Great Sankey
Warrington
WA5 3LW

Tel: (0925) 234000

Severn–Trent Water plc
2297 Coventry Road
Sheldon
Birmingham
B26 3PU

Tel: (021) 722 4000

Southern Water plc
Southern House
Yeoman Road
Worthing
BN13 3NX

Tel: (0903) 264444

South West Water plc
Peninsula House
Rydon Lane
Exeter
EX2 7HR

Tel: (0392) 219666

Thames Water plc
Nugent House
Vastern Road
Reading
RG1 8DB

Tel: (0734) 593333

Welsh Water plc
Plas-y-Ffynnon
Cambrian Way
Brecon
Powys
LD3 7HP

Tel: (0874) 623181

Wessex Water plc
Wessex House
Passage Street
Bristol
BS2 0JQ

Tel: (0272) 29061

Yorkshire Water plc
Broadacre House
Vicar Lane
Bradford
BD1 5PU

Tel: (0274) 306063

FORESTS & WATER GUIDELINES

HYDRO-ELECTRICITY GENERATORS

Scotland

Scottish Hydro-Electric
16 Rothesay Terrace
Edinburgh
EH3 7SE

Tel: (031) 225 1361

Scottish Power plc
Cathcart House
Spean Street
Glasgow
G44 4BE

Tel: (041) 637 7177

England and Wales

National Power plc
Dolgarrog Power Station
Dolgarrog
Conwy
Gwynedd
LL32 8QE

Tel: (0492) 69811

National Power plc
Mary Tavey Power Station
nr Tavistock
Devon
PL19 9PR

Tel: (0822) 810248

Power-Gen plc
Rheidol Power Station
Cwm Rheidol
nr Aberystwyth
Dyfed
SY23 2NE

OTHER ADVICE

Advice on fisheries is obtainable from a number of bodies including:

Scotland

Association of Scottish District Salmon Fishery Boards
The Stables
Cargill
Perth
PH2 6DS

Tel: (0250) 883365

Freshwater Fisheries Laboratory
Faskally
Pitlochry
Perthshire
PH16 5LB

Tel: (0796) 472060

England & Wales

Institute of Freshwater Ecology
Windermere Laboratory
Ambleside
Cumbria
LA22 0LP

Tel: (05394) 42468

MAFF Fisheries Laboratory
Pakefield Road
Lowestoft
Suffolk
NR33 0HT

Tel: (0502) 562244

The National Rivers Authority
(see local office address under WATER REGULATORY AUTHORITIES)

Water Research Centre
Medmenham Laboratory
Henley Road
Medmenham
PO Box 16
Marlow
Bucks
SL7 2HD

Tel: (0491) 571531

Advice on wildlife is obtainable from a number of organisations inlcuding:

Scotland

Scottish Natural Heritage
12 Hope Terrace
Edinburgh
EH9 2AS

Tel: (031) 447 4784

Scottish Wildlife Trust
16 Cramond Glebe Road
Edinburgh
EH4 6NS

Tel: (031) 312 7765

England & Wales

Countryside Council for Wales
Cyngor Cefn Gwlad Cymru
Plas Penrhos
Ffordd Penrhos
Bangor
Gwynedd
LL57 2LQ

Tel: (0248) 370444

English Nature
Northminster House
Peterborough
PE1 1UA

Tel: (0733) 340345

Royal Society for Nature Conservation
The Green
Witham Park
Waterside South
Lincoln
LN5 7JR

Tel: (0522) 544400

Advice on forest management is available from:

The local Forestry Authority conservancy office.

The Institute of Chartered Foresters
7a St Colme Street
Edinburgh
EH3 7HR

Timber Growers UK Ltd
5 Dublin Street Lane South
Edinburgh
EH1 3PX

APPENDIX II: THE WATER INDUSTRY AND WATER LEGISLATION

The water industry

The Water Act (1989) established the National Rivers Authority (NRA) and water utility companies in England and Wales. The NRA is charged with regulating water quality and resources, fisheries, recreation and amenity (on water and land within its control) and flood defence. The NRA's powers have been consolidated into the Water Resources Act (1991). The water utility companies provide public water supply, sewerage and sewage treatment. In some parts of the country water is supplied by water distribution companies.

In Scotland the 12 regional and Islands councils are responsible for water supply and sewerage, under the Water (Scotland) Act (1980) and the Sewerage (Scotland) Act (1968) respectively. The prevention of water pollution is the responsibility of the river purification authorities – the three Islands councils for the Islands, and the seven river purification boards on the mainland.

Public water supply

Under the Water Industry Act (1991), and preceding legislation, all water undertakers are required to supply wholesome water; standards of wholesomeness are prescribed by regulations under the Act.

These regulations are the Water Supply (Water Quality) Regulations (1989) (S.I. 1147), the Water Supply (Water Quality) (Amendment) Regulations (1989) (S.I. 1384), and the Water Supply (Water Quality) (Amendment) Regulations (1990), (S.I. 1837), and parallel regulations for Scotland. They prescribe bacteriological, chemical and aesthetic standards which incorporate and, in some instances, go beyond those set out in the EC directive relating to the quality of water intended for human consumption (80/778/EC).

Another relevant EC directive is that concerning the quality required of surface water intended for the abstraction of drinking water in member states (75/440/EC). It is implemented *inter alia* via the Surface Waters (Classification) Regulations (1989) (S.I. 1148), and by parallel regulations for Scotland. It is expected that the standards will become statutory water quality objectives for the NRA when so directed by the Secretary of State.

Control of water pollution

The Water Resources Act (1991) gives powers to the NRA, as does the Control of Pollution Act (1974) to RPAs, in respect of water pollution. It is an offence to cause or knowingly permit the entry of poisonous, noxious or polluting material into any inland water (lakes, lochs and watercourses), specified underground waters or tidal waters within the three nautical miles limit. Any discharge of sewage or trade effluent requires the consent of the NRA or RPAs, except that discharges by Islands councils need the consent of the Secretary of State for Scotland. There are powers available to make regulations requiring people storing polluting substances to take precautions to prevent their entry into water. There are also provisions for regulations to prohibit or restrict particular activities in designated protection zones, e.g. around a water supply source or important aquatic habitat.

The Food and Environment Protection Act (1985) requires prior consultation with the NRA or RPAs before the use of herbicides or pesticides in or near water, and before the aerial application of chemicals. Where there is a risk of contamination of water, herbicides which have been cleared for use in or near watercourses and lakes should be used. These are listed in the MAFF publication *Guidelines for the use of herbicides on weeds in or near watercourses and lakes* (1985).

Standards for metals and organic substances are increasingly being set in relation to EC directives, in particular those on *Pollution caused by certain dangerous substances discharged into the aquatic environment of the Community (76/464/EC)* and its daughter directives, and *Quality of fresh waters needing protection or improvement in order to support fish life (78/659/EC)*.

Fisheries

Freshwater fisheries are capable of legal ownership and anybody who negligently damages a fishery could be liable in damages to the owner or any other person having an interest in the fishery. Fisheries may be fished by their owners or let out to others and may be managed by individuals, companies, angling clubs or associations.

In England and Wales, under the Water Resources Act (1991) and the Salmon and Freshwater Fisheries Act (1975), the National Rivers Authority has a duty to maintain, improve and develop salmon, trout, freshwater and eel fisheries. The 1975 Act provides protection for fisheries and, for example, makes it an offence to render any water containing fish, or any tributary of that water, poisonous or injurious to fish, their spawning grounds, fish spawn or the food of fish.

In Scotland, under the Salmon Act (1986), local district salmon fisheries boards have powers to protect and improve salmon fisheries (but boards have not been set up in all districts). There is no statutory local administration for trout or other freshwater fisheries but, as in England and Wales, the legislation provides protection for all fisheries and, for example, prohibits the killing of fish with noxious substances, obstructing the passage of salmon, and damage to spawning gravel (Salmon Fisheries (Scotland) Act (1968) and Salmon and Freshwater Fisheries (Protection) (Scotland) Act (1951)). The right to fish for trout and other freshwater fish belongs to the owner of the riparian land, but salmon and sea trout fisheries are a separate heritable estate and may have a different owner.

Abstractions

In England and Wales the abstraction of water from rivers and underground waters requires a licence from the NRA under the Water Resources Act (1991). There are some exceptions, including water for fire-fighting. The impoundment of water also requires a licence, and in this case it includes storage for fire-fighting purposes when this is provided by a barrier constructed across a stream. The Salmon and Freshwater Fisheries Act (1975) requires the inclusion of a fish pass in any new dam or weir frequented by salmon or migratory trout.

In Scotland the right to abstract water from surface and underground sources is generally founded in the common law. There are, however, several statutes which govern abstractions for specific purposes.

These are:

1. Water (Scotland) Act (1980) – public water supplies.
2. Electricity Act (1989) – water for hydro-electric development.
3. Natural Heritage (Scotland) Act (1991) – covering all abstractions for irrigation, including spray irrigation, by commercially-based agriculture and horticulture.
4. Acts of Parliament (usually Private Acts) – water for specific purposes. Such acts are now rare as water for industrial purposes is usually obtained through the agency of the statutory authority.

Land drainage

In England and Wales the NRA, through its regional offices, exercises general supervision over all matters relating to land drainage and flood alleviation under the provisions of the Water Resources Act (1991). However, the Authority has no regulatory powers in respect of field drainage. It has permissive powers to undertake improvement and maintenance work on main rivers, designated by MAFF, and on sea defences in relation to the prevention of flooding. Internal drainage boards have drainage powers on small watercourses within their designated districts (mostly in low-lying agricultural areas), while local authorities have similar permissive powers on other non-main rivers.

A consent from the NRA is required before any structure affecting water flow in a main river is erected or modified. Byelaws give detailed provisions for controlling many activities in or beside main rivers – including construction work in or over the river and on the banks, the tipping of matter, removal of debris, management of flood banks, planting of trees and access to the river. Consent is also required from the NRA, or an internal drainage board where one exists, before the erection or modification of any dam, weir or other obstruction to flow in any watercourse. Likewise, no culvert likely to affect the flow of any watercourse should be erected or modified without consent. It is also an offence to obstruct the passage of migratory fish.

In Scotland, land drainage and the prevention of flooding of agricultural land are the responsibility of individual proprietors. The Scottish Office Agriculture and Fisheries Department administers grant-aided large-scale arterial drainage schemes involving co-operation by several proprietors. The regional councils have discretionary powers with regard to flood prevention on non-agricultural land, and protection of the coast from erosion and encroachment by the sea.

Conservation

In England and Wales, the Water Resources Act (1991) (Sections 16 to 18) makes statutory demands on the NRA and internal drainage boards. It requires that, when formulating or considering any proposals relating to the discharge of any of their functions, and as far as is consistent with their other duties, they should further the conservation and enhancement of natural beauty and the conservation of flora, fauna and geological and physiographical features of special interest. They must also have regard to the desirability of protecting buildings and other objects of archaeological, architectural and historic interest.

Comparable statutory duties do not apply in Scotland, but in 1982 The Scottish Office encouraged the water and sewerage authorities and the river purification authorities to carry out their functions in accordance with the spirit of the relevant statutory provisions then current in England and Wales. In addition, The Scottish Office has now published a voluntary Code of practice on conservation, access and recreation, which gives more detailed guidance for these authorities. The code has also been commended to a wider audience of interested parties in Scotland.

Many sites of special scientific interest and other protected areas involve rivers and lakes. The Wildlife and Countryside Act (1981) also has schedules of plants and animals, including aquatic species, afforded special protection.

APPENDIX III: THE FORESTRY INDUSTRY

Britain's area of productive woodland is more than two million hectares, producing over six million tonnes of timber each year. Output is expected to rise by 70% over the next 20 years as post-war plantations come into production. Government policy is that the national area of forests and farm woodlands should continue to increase.

The Forestry Commission is the government department in Great Britain responsible for forestry. The Forestry Act (1967) gives the Commission the general duties of: promoting the interests of forestry; the establishment and maintenance of adequate reserves of growing trees; the production and supply of timber; and the development of the recreational potential of the forests it manages. The Act requires that the Commission endeavours to seek a balance between its forestry functions and the conservation of flora, fauna and geological or physiographic features of special interest. **Forest Enterprise is the part of the Forestry Commission** which manages the national forest estate. Its operations include tree planting, estate management, civil and mechanical engineering, and the harvesting and marketing of timber. **The Forestry Authority is the part of the Forestry Commission** which provides grant aid to private forestry and regulates tree felling. It carries out research and exercises statutory controls to prevent the spread of tree pests and diseases. The Forestry Commission headquarters are at 231 Corstorphine Road, Edinburgh, EH12 7AT.

The Forestry Authority's research work on water-related matters is carried out at the research stations at Alice Holt Lodge, near Farnham, Surrey and at the Northern Research Station, near Roslin, Midlothian.

About *1.3 million hectares* of productive woodland is in private ownership. Private forestry is represented by **Timber Growers United Kingdom (TGUK).** Its membership includes owners with a few hectares of woodland, medium and large estates, personal and corporate investors, trusts and institutions. A number of local authorities, national park authorities and water authorities are forest and woodland owners. Many of the new forests established by personal and institutional investors are managed by professional forestry companies.

Timber may be harvested by a grower's employees, or sold standing to timber merchants. Wood-using industries may obtain supplies direct from the growers or from merchants. Both growers and merchants rely on contractors to carry out a wide range of forest work, in addition to their own directly employed workers.

The **Institute of Chartered Foresters (ICF)** represents and regulates the forestry profession in the U.K. Its members, as chartered foresters, possess the range of skills and knowledge required for the professional management of all forests and woodlands. A list of members in consultancy practice is available from the ICF.

GLOSSARY

Acid
An acid is a compound capable of transferring a hydrogen ion in solution and this definition is applicable throughout these guidelines.

Acidification
A continuing loss of acid neutralising capacity manifested by increasing hydrogen ion concentrations and/or declining alkalinity. May be applied to a catchment or to waters draining that catchment.

Alkalinity (acid-neutralising capacity)
and
Acidity (base-neutralising capacity)
A sample of pure water brought to equilibrium with air containing carbon dioxide at its normal partial pressure has a pH of around 5.6. Waters having a pH greater than this are said to possess **alkalinity** or **acid-neutralising capacity**, mainly in the form of bicarbonate and carbonate ions which react with added acid (hydrogen ions).

Waters having pH values less than 5.6 are said to possess **acidity** or **base-neutralising capacity.**

Note that 'acid-neutralising capacity' is synonymous with 'alkalinity'. However, the former expression can be used in a more general sense to refer to the ability of a *catchment* (not just a water body) to neutralise acid inputs.

Alevins
Newly hatched fish with a yolk sack, which live in the 'redds', or spawning gravel.

Aluminium (labile)
Aluminium is present in water in many different forms or **species,** the total concentration of which constitutes the variable most often measured. **Labile** aluminium is an operationally-defined, fast-reacting fraction of that total, believed to provide a better measure of aluminium toxicity.

Rapid passage of water samples through a cation-exchange resin column is often used to provide the operational separation of labile and non-labile fractions, the fast-reacting (i.e. labile) species being removed by the resin. Because the separation is empirical, the results obtained are likely to be sensitive even to apparently minor differences in the measurement procedures used.

Base (alkali)
A base used to be regarded solely as a chemical species capable of accepting a proton from another substance and this definition is applicable throughout this report. However, strictly speaking, a base is now defined more generally as a chemical entity capable of donating a pair of electrons (see Acid, above).

Base flow
Sustained runoff consisting largely of groundwater.

Brash
The residue of branches and tops, sometimes called 'lop and top', left on site following harvesting.

Brash mat
A mat of brash placed in rows on which harvesting and timber extraction vehicles run to reduce soil damage.

Brash thatching
The use of brash to bind together the road surface in the area receiving heavy use by forwarders stacking timber at the roadside.

Buffer area
An area which protects the watercourse from pollutants and sediment off the adjacent land. The buffer area will usually include the riparian zone and may extend into the adjacent land.

Buffer capacity (buffer intensity; buffer index)
Measure of the resistance of a solution to pH change. All well buffered solutions contain a weak acid. In waters of around neutral pH the bicarbonate/carbonate system constitutes the weak acid buffer. In acid waters, other acids (e.g. organic) or aluminium complexes act as buffers.

Bulk precipitation
The total input of wet and dry deposition measured by a horizontal gauge.

Bunded tanks
Tanks for fuels, oils or other chemicals which are protected by containment within a fluid resistant dyke.

Cable–crane extraction
A timber extraction method which hoists logs clear of the ground and carries them to the roadside on an aerial cable stretched between standing trees.

Continuous-acting mounders
Tractor-mounted or trailed machines which produce mounds by digging or scraping the soil surface as the machine moves over the site.

Coupe
An area of woodland. A felling coupe is an area of woodland designated for felling.

Critical loads
The most commonly used definition is *'the highest deposition of acidifying compounds that will not cause chemical changes leading to long-term "harmful" effects on ecosystem structure and function'.* This definition is ambiguous due to the subjectivity required to judge a long-term harmful effect. So usually the 'harmful' is omitted and the definition is improved.

Critical load exceedance
The critical load for an individual site is exceeded when the first change in the aquatic ecosystem that can be related to acid deposition occurs.

Cross drains/cut-off drains
Open ditches aligned slightly off the contour with a bed gradient not exceeding 2°. Their primary function is to control the build up of potentially eroding water movement down cultivation channels following high intensity rainfall. Drains can affect the water table on some wet peaty soils.

Cultivation channels
Any linear cultivation feature including plough furrows, mole drains and scarifier trenches.

Dry deposition
Pollutants reaching the ground in particulate form or as aerosols or gases, i.e. excluding the indirect input in aqueous solution or suspension (wet deposition).

Episode
An intensive, short-term surge of stream water or atmospheric pollution characterised by rapidly changing chemical composition. Episodes are usually associated with heavy rainfall or rapid snow melt, or other changes in meteorological conditions.

Excavator mounding
The technique provides small 30–50 cm high mounds on which to plant trees using soil excavated from an adjacent drain. The drain spacing is controlled by the maximum reach of the machine.

Forwarders
Tractors which extract timber lifted entirely clear of the ground. The timber is carried on a linked trailer or integral platform.

Groundwater
Water stored in the soil and rock both above and below the water table.

Headwater source areas
Wet flushes, bogs or springs at the head of first-order streams.

Horizon
A layer of soil which may differ in colour, texture and composition from other layers lying above or below.

Kick sample
A standard method for sampling benthic macroinvertebrates in running water. The analyst holds a net vertically on the river bed downstream of the foot, and the toe or heel of the boot is used to disturb the substratum, releasing material to be caught in the net.

Metalliferous
Road construction material containing metals which may leach out in solution and which may lead to the pollution of watercourses. Some materials to avoid are mining spoil, opencast spoil, and oil shale.

Mineralization
Production of inorganic ions in the soil by the oxidation of organic compounds.

Moling
A cultivation method similar to ripping but with an additional device to open a conduit within the soil along which water may flow to cross drains.

Nitrate leakage
Quantities of nitrate can leak into watercourses when the rate of supply of nitrate exceeds its utilisation.

Nitrification
Oxidation of ammonia to nitrites and of nitrites to nitrates, as by action of bacteria.

Occult deposition
The turbulent transfer of cloud, mist or fog droplets, containing large concentrations of pollutants, to vegetation.

Overland flow (return flow)
Water passing rapidly over or through the surface layer of soil.

Peat
An organic substrate formed of partially decomposed plant material. The Forestry Commission defines peats as having a depth greater than 45 cm.

Peaty gley
A wet, imperfectly draining soil type in which a peat layer less than 45 cm thick overlies a mineral soil which is periodically water-logged.

pH
A logarithmic index for the hydrogen ion concentration in an aqueous solution. Used as a measure of acidity of a solution. Given by $pH - \log_{10}[H^+]$ where $[H^+]$ is taken as the hydrogen ion concentration. A pH below 7 denotes acidity and one above 7 denotes alkalinity. pH measurements give no indication of how the hydrogen ion concentration of a water came about, or how it will change with the addition or removal of other substances.

Piped crossing
A length of corrugated plastic pipe of about 1m diameter, placed in a watercourse as a temporary culvert to enable vehicles to cross during forest operations. The piped crossing often carries a brash mat.

Pipeflow
Water moving rapidly through the soil in natural cracks and channels.

Podzol
A soil type in which there is a surface layer of acid humus and below this a severely leached mineral layer; typically found under coniferous forest and heathland.

Pollution climate
A phrase used to describe the mixtures of pollutants that occur on a regional scale.

Precipitation
Rain, snow, hail or sleet. Occult precipitation is the accumulation of water on surfaces from cloud, fog or mist.

Ripping
A cultivation method in which a tine pulled behind a tractor disturbs the soil.

Runoff
The gravity flow of water in open channels.

Salmonids
Fish belonging to the family Salmonidae, including salmon, brown trout, sea trout, grayling, powan and charr.

Scarifiers
A tractor-mounted or trailed cultivation machine which scrapes away a shallow layer of harvesting debris or ground vegetation to provide a clear weed-free planting site.

Skidding
The extraction of timber using a tractor to lift one end of the log clear of the ground with the other end dragging on the ground.

Soil acidification
This may be defined in two ways: 1. a process causing a decrease in soil solution pH; 2. a process which reduces the acid-neutralising capacity of the soil. The former definition is the one adopted in these guidelines.

Spaced furrow ploughing
A tractor-mounted or trailed plough used to create warm, weed-free, aerated planting sites on the plough ridge. Ridges are formed in ribbons spaced at distances suitable for the growth of trees.

Spoil
The soil material resulting from the excavation of open drains, or material excavated from silt traps.

Stemflow
Water that reaches the ground by flowing down the surfaces of stems.

Throughfall
Water that: 1. drips from plant surfaces (crown drip); and 2. falls uninterrupted through gaps in the canopy (direct penetration).

Throughflow
Water passing slowly through the subsurface layers of soil.

Water regulatory authority
In Scotland, the local river purification authority (RPA) (the seven river purification boards (RPBs) on the mainland, and the three Islands councils for their area) and in England and Wales, the National Rivers Authority.

Water undertaker
In Scotland, the regional or Islands councils, and in England and Wales, the water utility companies.

Wet deposition
Pollutants reaching the ground in rain and snow or as occult deposition.

ACKNOWLEDGEMENTS

This edition of the guidelines was updated by a core group made up of:

A H A Scott (Chairman), Forestry Practice Division, The Forestry Authority
J G S Gill, Forestry Practice Division, The Forestry Authority
P H Freer-Smith, Research Division, The Forestry Authority
A S Gee, National Rivers Authority, Welsh Region
A P Donald, Scottish Office Environment Department
W T Welsh, Solway River Purification Board
B R S Morrison, Scottish Office Agriculture & Fisheries Department
D Ray (Secretary), Research Division, The Forestry Authority

Comments and advice were received from the following:

Countryside Commission
Countryside Council for Wales
Department of Agriculture for Northern Ireland
Department of the Environment
English Nature
HM Inspectorate of Pollution
Institute of Freshwater Ecology
Institute of Hydrology
Institute of Terrestrial Ecology
Macaulay Land Use Research Institute
Scottish Natural Heritage
The Scottish Office Agriculture & Fisheries Department
The Scottish Office Agriculture & Fisheries Department, Freshwater Fisheries Laboratory, Pitlochry
The Scottish Office Central Research Unit
Water Research Centre
Welsh Office

University of Aberdeen, Department of Plant & Soil Science
University of Cambridge, Department of Zoology
University College London, Environmental Change Research Centre
University of Newcastle upon Tyne, Department of Geography
University of Wales, School of Pure & Applied Biology

David Goss Associates
Forest Enterprise
Fountain Forestry Ltd
Institute of Chartered Foresters
National Farmers' Union of Scotland
Scottish Landowners' Federation
Scottish Woodlands Ltd
Tilhill Economic Forestry Ltd
Timber Growers UK Ltd

National Rivers Authority
Clyde River Purification Board
Forth River Purification Board
Highland River Purification Board
North East River Purification Board
Solway River Purification Board
Tay River Purification Board
Tweed River Purification Board

Scottish Hydro Electric plc
Scottish Power plc

Anglian Water plc
South West Water plc
Welsh Water plc
Wessex Water plc
Yorkshire Water Enterprises

Confederation of Scottish Local Authorities (COSLA)
Borders Regional Council
Central Regional Council
Dumfries & Galloway Regional Council
Grampian Regional Council
Strathclyde Regional Council
Central Scotland Water Development Board

The Institute of Biology
Royal Society for Nature Conservation
Royal Society for the Protection of Birds
Scottish Countryside Activities Council
Wildlife Link

The Association of District Salmon Fisheries Boards
The Atlantic Salmon Trust
The British Trout Association
The Institute of Fisheries Management
The Institute of Fisheries Management, Scottish Branch
The River Halladale District Salmon Fisheries Board
The Salmon & Trout Association
The Tweed Foundation

The critical loads map was provided by the Department of the Environment; and the assistance of K Bull of ITE, Monkswood is gratefully acknowledged. The working group is also grateful to D G Pyatt, G S Patterson and T R Nisbet of the Forestry Authority Research Division for drafting sections of this revision.

BIBLIOGRAPHY

Department of the Environment: *Acid rain: critical and target loads maps for the United Kingdom*, 1993.

Department of the Environment and Forestry Commission: *Forests and surface water acidification*, 1991.

Forestry Commission: Handbook 6, *Forestry practice*, ed B G Hibberd, 1991.

Forestry Commission: Bulletin 86, *Forests and surface water acidification.* T R Nisbet, 1990.

Forestry Commission: Bulletin 107, *A manual of good practice for the use of sewage sludge in forestry.* R Wolstenholme, J Dutch, A J Moffat, C D Bayes and C M A Taylor, 1992.

Forestry Commission: Research Information Note 148, *Liming to alleviate surface water acidity*, 1989.

Forestry Commission: Research Information Note 196, *Forest drainage*, 1990.

Forestry Commission: Occasional Paper 21, *Provisional code of practice for the use of pesticides in forestry*, 1989.

Forestry Commission: *Forest landscape design guidelines*, 1989.

Forestry Commission: *Forest nature conservation guidelines*, 1990.

Forestry Commission: *Lowland landscape design guidelines*, 1992.

Forestry Commission: Woodland Grant Scheme. 1993.

Forestry Commission: *Tree felling – licences and permissions*, 1993.

Forestry Commission: *Consultation for grant and felling applications*, 1993.

Ministry of Agriculture, Fisheries and Food: *Pesticides: code of practice for the safe use of pesticides in farms and holdings*, 1990.

Ministry of Agriculture, Fisheries and Food: *Guidelines for the use of herbicides on weeds in or near watercourses and lakes*, 1985.